The Scienc of Pain, TENS & Acupuncture

Dr. Richard Cheng
M.D. Ph.D.

The SCIENCE

of

PAIN, TENS

&

ACUPUNCTURE

Author: Dr. Richard S. S. Cheng M.D. Ph.D.

Edited by: Winnie Cheng, Weyland Cheng, Ricky Cheng & Irene Schwerk

The Science of Pain, TENS** and Acupuncture

I. **Pain and Pain Mechanism** 6
II. **Endorphins** (Endogenous Morphines) 14
III. **History of Acupuncture in Chinese Medicine** 27
IV. **The Classical Theory of Acupuncture** 29
V. **New Meridian Theory** 31
VI. **History of Electrotherapy** 41
VII. **Mechanism of Acupuncture and TENS**** 42
VIII. **Fractal Stimulation** 48
IX. **Pain Syndromes and Treatment Methods** 58
X. **Contraindications and Precautions for Acupuncture, EA* and TENS**** 67
XI. **Analysis of Acupuncture Points** 68
XII. **Techniques & Clinical Approaches for Acupuncture, EA* and TENS**** 71
XIII. **Clinical Treatment of:**

1. **Headache** 73
2. **Musculoskeletal Disorders of :**
   a) Neck 84
   b) Shoulder 89
   c) Elbow 93
   d) Forearm, wrist, hands and fingers 96
   e) Upper back 104
   f) Chest and abdomen 105
   g) Lower back 108
   h) Hips 118
   i) Knee 125
   j) Ankle, feet and toes 132

   k) Lupus 138
   l) Osteoporosis 139
   m) Rheumatoid Arthritis 140

3. **Cardiovascular Disorders**
   a) Hypertension 141
   b) Hypotension 142

4. **Gastroenterology**
   a) Anorexia 143
   b) Colitis, enterocolitis and Crohn's disease 144
   c) Constipation and diarrhea 145
   d) Hepatitis 146
   e) Ileitis and paralytic ileus 147
   f) Abdominal Pain/ Irritable bowel Syndrome 148
   g) Obesity 149

5 **Genitourology**
a) Impotence 150
b) Nephritis 151
c) Prostatitis and Prostate Hypertrophy 152
d) Renal Colic (kidney stones) 153

6. **Immune System Problems** 154
Asthma, AIDS, Eczema, Hay Fever, Lupus, Rheumatoid Arthritis, Psoriasis, Crohn's disease, Scleroderma etc.

7. **Neurological Problems**
a) Alzheimer's disease 155
b) Diabetic neuropathy, ischemic leg pain 156
c) Insomnia 157
d) Meniere's disease 158
e) Minor depression and nervousness 159
f) Post herpetic neuralgia – corresponding dermatome
g) Post surgical neuralgia – corresponding dermatome
h) Restless leg syndrome and leg cramps – (7 b) 156
i) Sympathetic reflex dystrophy - corresponding dermatome & muscles
j) Tinnitus 160
k) Trigeminal neuralgia 161

8. **Obstetrics – Gynecology**
a) Fibroids 162
b) Menopause Syndrome and Menstrual Pain 163

9. **Children's (Pediatric) Problems**
a) Growing Pains – (7b) 156
b) Hyperactive and Learning/ Hearing problems 164
c) Others: Asthma and Hay Fever (10b) 165
Neurodermatitis (12b) 169
Juvenile Rheumatoid Arthritis (2m) 140

10. **Respiratory System**
a) Allergic Sinusitis and Hay Fever 165
b) Asthma, COPD and Chronic Cough 166
c) Common Cold, Flu 167

11. **Rheumatology**
a) Immune system (6) 154
b) Lupus (2 k) 138
c) Rheumatoid arthritis (2m) 140

12. **Skin**
a) Acne 168
b) Allergy, Hives, Eczema, neurodermatitis 169
c) Alopecia 154
d) Psoriasis (6) 154

13. **TENS** for Facelifting** 170
(Muscle Tightening and Weight Reduction)

14. **Medical Qi–Gong Meditation** 172

15. **Microcosmic Orbit** 174

Appendix I: 14 Meridians and the Ear Points 175

Appendix II: Illnesses Treated by Acupuncture 190

References 195

* **EA** – Electroacupuncture
** **TENS** – Trancutaneous Electrical Nerve Stimulation

I. PAIN AND PAIN MECHANISM

In human experience, pain is one of the most debilitating symptoms known to man. Pain can be roughly categorized into two classes – acute and chronic. In the acute stage, pain serves as an alarm system warning us that something biologically harmful is happening to our body, while in the chronic stages, pain often causes drastic emotional and physical stress. It has been estimated that chronic pain costs the American people alone more than $500 billion per year. This economic burden and the psychological disorder induced by pain create a serious national and worldwide health problem. The mechanism of pain, therefore, has been investigated by numerous professionals in the fields of psychology, biology and medicine. The contributions made by these people toward understanding the mechanisms of pain have often given rise to conflicting observations – the result being an overabundance of interpretations of pain mechanisms. In spite of this, a general understanding of pain perception and transmission has prompted a great deal of research.

Psychologically and physiologically, scientists try to classify pain and its pathways into three dimensions:

(i) **Sensory-discriminative** –a sensory type of pain which is subserved, at least in part, by the neospinothalamic projection to the ventrobasal thalamus and the somatosensory cortex, via the anterolateral thalamic tract (ALT).

(ii) **Motivational-affective** – an emotional type of pain which is believed to involve the brainstem reticular formation and the limbic system which receive projections from the spinoreticular and paleospinothalamic components of the anterolateral somatosensory pathway (via ALT).

(iii) **Cognitive-evaluative** – it is known that cultural values, anxiety, attention and suggestions have a profound effect on pain experience which may involve cortical control over the sensory discriminative and motivational-affective dimension of pain.

Pain, however, often arises from intense stimuli that cause tissue damage. It is believed that this tissue damage leads to an accumulation of certain chemicals, such as histamine, bradykinin, prostaglandin E, etc., which induce the discharge of afferent nerve impulses. Lindahl (1974), with the use of a pH microelectrode in the tissues, has found that pH is very acidic in tissues where the bare nerve endings are situated. Thus he has proposed that a pH change is the final chemical change that causes pain. Pain, however, is also elicited by the mechanical deformation of simple free nerve endings.

Many hypotheses have been proposed to explain the mechanisms of pain. Some of them are still considered to be useful in explaining certain physiological phenomena:

(1) The specific modality theory hypothesizes that pain is detected by high threshold free nerve endings and transmitted through small afferent fibres (like A-delta or C fibres) and the spinothalamic tracts to the focal pain centres in the brain.

(2) The pattern theory proposes that all sensory inputs are transmitted through the same pathway, but are interpreted differentially according to the spatiotemporal pattern of the sensory impulses at the higher nervous centre.

(3) The gate control theory, proposed by Melzack and Wall (1965), suggests that a gate for the control of pain exists at the spinal cord level. Pain can be facilitated or inhibited according to the quantity of small afferent and large afferent inputs. However, the facilitation of pain by primary afferent hyperpolarization (PAH) does not exist; only primary afferent depolarization (PAD) exists in large fibres. Thus the gate control theory as described by Melzack and Wall (1965) requires a neurophysiological revision. Despite the controversy and conflicting evidence, the gate control theory gave a stimulating idea on the mechanism of pain and widened the field of pain research. Melzack (1973) later stressed that this gate is also subjected to the influence of higher brain centres.

1. PRIMARY AFFERENT NEURONS

There is convincing physiological evidence that specialization exists within the somaesthetic system. The skin receptors have specialized physiological properties. Free nerve-endings, however, may all look alike despite their highly specialized properties. Hensel and Andres (1974) found a correlation between the structure of the receptor and the afferent discharge. They marked a sensitive spot on the skin, excised it and examined it with an electron microscope. They claimed to have found distinctive structures for the free nerve endings.

The special properties of different primary afferent neurons have been extensively studied. They were briefly summarized as follows:

A. MECHANO-SENSITIVE NEURONS

(i) **High threshold A-delta mechanoreceptive afferents** –
These neurons respond only to intense mechanical stimuli and have been found in the skin of cat and monkey facial skin. Conduction velocities usually range from 15-25 m/sec, and receptive fields on limbs are from 1-8 sq.cm., with more than one sensitive spot.

(ii) **Low sensitivity and moderate pressure A-delta mechanoreceptive afferents** - These neurons are found in the same area as (i) but respond to tissue-threatening and tissue-damaging stimuli.

(iii) **High threshold C mechanoreceptive afferents** -
These are superficial unmyelinated neurons innervating the extremities of cats and monkeys. They have small receptive fields of small C fibre population. Some of these neurons in cats also respond to cold temperatures.

(iv) **Low threshold myelinated (AB and A-delta) mechanoreceptive afferents** – These neurons are sensitive to weak mechanical stimuli and show no differential response to noxious mechanical stimuli. Most of these neurons function as position and/or velocity detectors on the skin. They synapse with the spinal cord interneurons that project to the dorsal column, spinocervicothalamic, spinothalamic and spinoreticular pathways. Electrical stimulation of AB fibres in the periphery or dorsal columns does not produce pain. However, some low threshold, slowly adapting AB mechanosensitive afferents converge on spinothalamic and trigeminothalamic neurons that most likely participate in the sensory – discriminative aspect of pain. Their activity may summate with nociceptive inputs terminating on these central neurons. They may play a role in pain modulation as their major effects are to suppress spontaneous activity and nociceptive responses of various types of spinal cord neurons. Electroacupuncture or transcutaneous-stimulation effect may be mediated by these slowly adapting low threshold A beta fibres and/or A-delta fibres.

(v) **Low threshold unmyelinated C mechanoreceptive afferents** – These neurons respond to weak, slowly moving mechanical stimuli. They comprise about 50 % of the C fibres in cats, 10% in monkeys and none in humans.

B. THERMOSENSITIVE NEURONS

(i) **"Cold" A-delta thermoreceptive afferents** – These neurons respond to small decreases in temperature (< 1°C) with receptive fields (single spots) less than 500 um in diameter. They may respond to steady-state temperature in the noxious heat range (45-52°C), and exhibit spontaneous activity in the 20–30°C range.

(ii) **"Warm" A-delta and C thermoreceptive afferents** – These neurons are similar to (i) and are sensitive to < 1°C changes in temperature, but warming increases and cooling suppresses their discharges. Regular neuronal discharges are seen at steady-state temperature in the range of 30–43°C. They comprise 50% of A-delta fibres in the face of the monkey.

(iii) **High threshold A-delta and C thermoreceptive** afferents – These fibres receive wide-dynamic range of stimuli and very few of them respond only to noxious heat or cold.

C. MECHANICAL-THERMOSENSITIVE NEURONS

(i) **A-delta heat nociceptive neurons** – These neurons can be found in limb and facial skin of monkeys and limb skin of cats. The receptive fields are usually less than 5 mm^2 and respond to a temperature range of 45–53°C as well as to non-noxious mechanical stimuli.

(ii) **C-polymodal nociceptive afferents** – These fibres constitute 80-90% of the primate C fibre population and respond to high threshold mechanical and thermal stimuli as well as irritant chemical stimuli.

Non-noxious mechanical (< 1g) and heat (> 38-40°C) stimuli also activate these neurons. Strong evidence indicates that C polymodal nociceptors are the important afferents that signal the presence of tissue damage.

(iii) **Mechanical cold A-delta and C nociceptive afferents** – Several reports indicate that these primary afferents respond only to noxious cold and high threshold mechanical stimuli.

2. SPINAL CORD

By using different techniques, such as suppressive silver staining and axonal transport, LaMotte (1997) has demonstrated that small fibres end in laminae I, II and III whereas large fibres are believed to end in laminae IV, V and VI. The primary afferent inputs to the dorsal horn have been studied by many researchers and are described in a review by Kerr and Casey (1978). Evidence has also indicated that the medical aspect of the tract of Lissauer exerts a facilitative effect on the sensory inputs and that the lateral part is inhibitory.

Rexed (1952) divided the dorsal gray into nine layers (laminae) according to the different sizes of the neurons. In general, Cervero et al (1976) have classified the dorsal horn neurons into three classes:

- **i.** **Class I** – excited by cutaneous mechanoreceptors, these neurons are located in lamina IV and some in V.
- **ii.** **Class II** – excited by both mechano- and nociceptors, these neurons are wide-dynamic range neurons which are located in lamina V.
- **iii.** **Class III** – respond only to noxious input. Some of them receive A-delta input and the others may be excited by both A-delta and C fibres. These neurons are located mostly in lamina I and occasionally in lamina II. Thus, they may represent another type of nociceptive processing as compared to class II neurons.

Kumazawa and Perl (1978) have reported that A-delta input is localized to neurons in the marginal zone while the C-fibre inputs synapse mostly in the substania gelatinosa. Visceral and somatic activities usually converge on single neurons in deeper layers such as laminae V, VI and VII of the dorsal horn. In conclusion, it appears that two populations of nociceptive neurons exist in the dorsal horn; the marginal neurons, which are mostly excited by high thresholds. By means of retrograde axoplasmic transport studies and recordings of antidromic spikes, neurons from layers I, IV, V, VI and VII show projections to the thalamus, and layer VIII cells show projections to the reticular formation.

3. ASCENDING PAIN-SIGNALLING PATHWAYS

Experimental evidence indicates that nociceptive transmission is by many pathways. A review shows that there are six pain-signalling systems; three lateral and three medial. The lateral group is comprised of: (1) the neo-spinothalamic tract, (2) the spinocervical tract and (3) the postsynapic tract of the dorsal column. The medial group is formed by (4) the paleo-spinothalamic tract, (5) the spinoreticular tract and (6) the diffuse polysynaptic-propriospinal tract. The medial group differs from the lateral group in having slower conduction velocity, cell bodies located more deeply in the spinal grey and different patterns of termination. Obviously, the two groups would exert different functions in pain-signalling systems. Moreover, all six pain-signalling pathways (both lateral and medial) have different anatomical routes ascending from the spinal cord to the brain and are controlled by different pain-inhibitory systems from the brain. The detailed studies of these pain-transmitting pathways are briefly reviewed as follows:

i. **Neo-spinothalamic Tract** (nSTT) – ascends from the ventral and ventrolateral regions of the spinal cord to the thalamus. In primates, 30% of the tract fibres respond to intense mechanical or thermal stimuli, 38% to hair movement, 21% to light pressure and 11% to deep, subcutaneous stimulation. It is estimated that 12.2% of ventrolateral cord fibres respond only to noxious stimuli, whereas 61% responded to both light tactile and painful stimuli carried by the neospinothalamic tract in primates and cats.

ii. **Spinocervical Tract** (SCT) – postsynaptic fibres ascend the dorsolateral spinal funiculus to the lateral cervical nucleus which projects to the somatosensory thalamus and adjacent areas and to the reticular formation through the medial lemniscus. Recording from 25 SCT cells which respond to light tactile stimuli, 22 (44%) increased discharge to intense mechanical stimulation; 6 of 16 cells were sensitive to noxious heat while one responded only to noxious stimulation. About 75% of the fibres ascend ipsilaterally to the lateral cervical nucleus and 25% go to the rostral dorsal column nucleus. The SCT is vestigial in some people as the lateral cervical nucleus was detected only in 9 out of 16 human spinal cords. In conclusion, a high percentage of the SCT fibres respond to peripheral A-delta and C fibre stimulation, showing the "wind-up" effect which may be related to slow pain.

iii. **Dorsal Column System** – It has been known for decades that the primary afferent fibres ascending to the dorsal column nuclei (DCN) carry only innocuous tactile and proprioceptive sensation. However Uddenberg (1968) showed that the dorsal column postsynaptic (DCPS) fibres transmit noxious messages. Recording from the cervical dorsal column of cats, he found that 79 out of 295 axons responded to small peripheral fibre (A-delta and C fibres) stimulation. Anguat-Petit, and Petit confirmed Uddenberg's finding by showing that 92 units (9.3% of the DC fibre) were DCPS axons of which 77.2% were sensitive to both gentle and noxious stimuli, 6.5% only to noxious mechanical stimuli and the rest only to light mechanical stimuli. Thus pain-signalling information may be carried by the DCPS to the dorsal column nuclei which project to various thalamic and collicular regions and to the zona incerta by the medial lemniscus, and eventually relay to the cortex.

The three lateral systems share several pain-signalling characteristics. Each has rapid conduction velocity, and carries noxious, thermal and light tactile sensations. They all originate from the dorsal horn of the spinal grey and project mainly to the lateral thalamus (ventrobasal, posterior and subthalamic complexes). However, qualitative and quantitative differences exist. For example, about 6.7% of the feline DCPS cells respond exclusively to noxious stimulation, as compared to 20% of the SCT. The inhibitory controls on these three systems can be achieved by stimulating many spinal cord and brain areas such as the contralateral, dorsolateral and ventromedial cord, the dorsal column, the mesencephalic tegmentum, central pontobulbar core and several cerebellar regions and specific cortical regions. The inhibitory control of SCT appears to be both cortical and subcortical because decerebration does not abolish SCT inhibition. The inhibitory effects are exerted mostly on noxious signals in the SCT system, leaving the light tactile responses unaltered. In contrast, cortical stimulation inhibits only the light tactile inputs, leaving the nociceptive responses (especially from lamina I) intact in the nSTT units. Inhibition of noxious inputs in the nSTT cells can be triggered by stimulating the nucleus raphe magnus and the caudal part of the nucleus gigantocellularis.

The three medial pain-signalling systems are:

(1) **The paleo-spinothalamic tract** – also ascends from the ventral and ventrolateral regions of the spinal cord and projects to the midline and intralaminar thalamic nuclei. Nociceptive signals have been detected in the midline-intralaminar nuclei.

(2) **The spinoreticular tract** – ascends from the ventral and ventrolateral regions of the spinal cord and projects to various sites of reticular formation and brain stem central grey. The spinal input to the reticular formation is diffuse, multisynaptic and interrelated to other sensory modalities.

(3) **Propriospinal tract** – nociceptions may be carried up and down the spinal cord along the ventral and lateral region of the spinal grey, forming synapses with other sensory systems or pathways. This tract is listed here for completeness.

II. ENDORPHINS (Endogenous Morphines)

1. Opiate Receptors

Biochemical evidence indicates the existence of multiple opiate receptors. There are at least three types of opiate receptors: u-receptors, k-receptors and d-receptors (others included sigma and epsilon). Each of these receptors possesses a specific ligand: morphine activates u-receptors for analgesia and euphoria, ketocycazocine activates k-receptors for feelings of sedation and benzomorphin derivative activates d-receptors for dysphoric and hallucinatory syndrome. It has been found that enkephalins act mainly at delta receptors, beta-endorphin at mu and dynorphin at kappa receptors.

2. Types of Endorphins

The detection of the opiate receptors aroused the search for endogenous opiates. It was predicted that opiate receptors and the "endogenous opiates" served as a lock-and-key system. It was hypothesized that since there is a lock, there must be a key to fit this lock. Successful results were obtained by several groups of workers; several endogenous ligands of the opiate receptors were detected in the brain, CSF, pituitary and peripheral tissues. These endogenous ligands, called endorphins (endogenous morphine) are peptides which have a similar effect to morphine in inducing analgesia, euphoria, relieving depression and affecting other important metabolic functions. Different types of endorphins have been detected in the pituitary, brain, spinal cord, CSF, blood plasma and peripheral tissues. There are at least four kinds of endorphins in the mammalian nervous system: Met-enkephalin, Leu- enkephalin, Beta-endorphin and dynorphin. These opioid peptides have a common NH2 terminus (Tyr-Gly-Gly-Phe-Met or Leu) but different COOH termini.

The following table summarizes the number of endorphins that have been detected:

Types of Endorphins

1. Mehionine- enkephalin (H-Tyr-Gly-Gly-Phe-Met-OH)
2. Leucine-enkephalin (H-Tyr-Gly-Gly-Phe-Met-OH)
3. Pro-opiocortin 31,000 or 37,000 daltons
   (Precursor for ACTH, beta-MSH and beta-LPH)

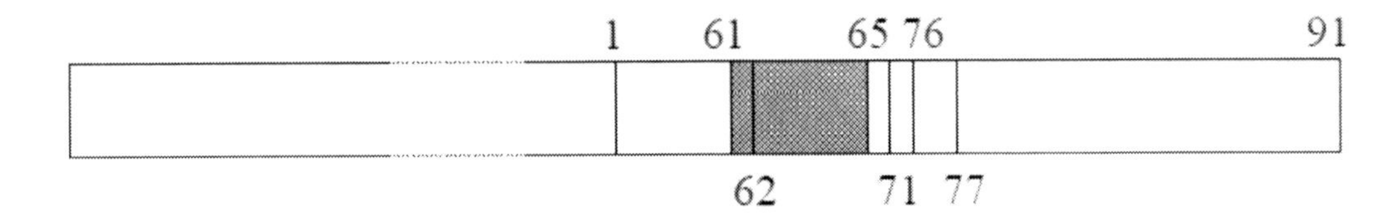

(Pro-opiocortin – ACTH, beta-MSH and beta-LPH)

Beta-lipotrophin (1–91)
Beta-endorphin (61–91)
Alpha-endorphin (61–76)
Gamma-endorphin (62–77)
Des-tyr-1-gamma-endorphin (62–71)
Met-enkephalin (61–65)

4. Dynorphin (17 amino acids) – in posterior pituitary and spinal cord.

Other Related Substances

1.) Endorphin releasing factor (H-Tyr-Arg-OH)
2.) Endogenous anti-opiate substance - cholecystokinin octapeptide

3. Location of Endorphins

(i). Enkephalins

Met-enkephalin and leu-enkephalin are two pentapeptides first identified in the brain extracts by Hughes and colleagues. They appear to be neurotransmitters in nerve terminals and mimic the effect of morphine by binding to opiate receptors. By comparing the localization of opiate receptors and enkephalins, it is found that they occur in similar places and correspond closely to the bodily functions known to be affected by opiate drugs. Although the concentration of met-enkephalin in the entire brain is generally three times higher than leu-enkephalin, the ratio varies in different brain areas and peripheral tissues.

Immunofluorescence shows that enkephalins in the brain are confined to neurons and are mostly located in the nerve endings, while 85% of the brain supporting cells (glia) show no trace of enkephalin fluorescence.

Indirect immunofluorescence revealed met-enkephalin immunoactive cell bodies in the te-, di-, mes- and rhombencephalon. In general, high concentration of enkephalins were found in the brain regions in the following order:

Striatum > Hypothalamus > Thalamus > Hippocampus >
Pons and Medulla > Cortex >>> Cerebellum

Enkephalins are also found in rat pituitary. The levels of met-enkephalin and leu-enkephalin are highest in the pars intermedia (7 and 4 p mole/mg respectively), medium in the pars anterior (0.51 and 0.36 p mole/mg respectively). In human CSF, enkephalin concentration is about 3.12- 3.25 p mole/ml.

In the spinal cord, enkephalins are located in the grey matter, especially laminae I, II and III which are packed with small neurons that interact with each other and impinge on the nerve endings of the sensory neurons. A chronic lesion of the primary afferents decreases the number of opiate receptors in the dorsal horn with no loss of enkephalin in cell bodies in the spinal cord. It is also shown that electrical stimulation, or K^+ provoke the release of substance P from slices of rat trigeminal nucleus and of rat substantia nigra, and that morphine and endorphins cause a concentration-dependent, stereospecific, naloxone-antagonizable reduction of its release. This suggests that met-enkephalins are found in neurons in the dorsal horn and that enkephalins act as presynaptic inhibitors in the spinal cord.

Enkephalins also appear in the peripheral tissues and digestive tract. Large amounts of enkephalins were found in the small intestine with the highest concentration occurring in the longitudinal muscle-myenteric plexus layer, and small quantities in the circular mucosal layer. Both met-enkephalin and leu-enkephalin are also found in the coeliac and superior cervical ganglia of rats, and in adrenal medullary gland cells.

(ii). Beta-endorphin

A "big" beta-lipotropin (LPH) or "big-big" beta-endorphin (pro-opiocortin) was found to be the precursor for ACTH and beta-LPH (beta-endorphin) with molecular weight of 37,000 daltons in human and 31,000 daltons in the rat pituitary and brain. The pro-opiocortin is broken down by peptidase to

ACTH, and beta-lipotropin is further cleaved into beta-endorphin. It is believed that the small amounts of alpha-endorphin, gamma-endorphin, des-Tyrl-gamma-endorphin are derived from beta-lipotropin. ACTH and beta-endorphins are stored in the same secretory granules of anterior pituitary. They are released together. Beta-endorphin is mainly used for pain relief and ACTH has anti-innflammatiory effect as ACTH stimulates cortisol synthesis and secretion by the adenal glands. Structural analysis of beta-LPH indicates that some of its sequence portions are related to secretin, human growth hormone (HGH), ACTH and proinsulin. These local sequence homologies, from an evolutionary point of view, reveal that the genome of beta-LPH may have arisen from the fusions of primordial genes for various small peptides, like secretin-glucagon, which are connecting peptides of proinsulin.

Large amounts of beta-endorphin and alpha-endorphin are found in the pars intermedia of the hypophysis, and in discrete cells of the adenohypophysis (pars distalis). These adenohypophyseal cells, containing the beta-endorphin and alpha-endorphin, often appear to be adjacent to blood vessels. The pars nervosa (neurohypophysis, posterior lobe) and the interlobular stroma of the pars intermedia contain no trace of these endorphins. Beta endorphin is also present in the brain and is not derived from the pituitary. This has been shown by the fact that hypophysectomy does not alter the level of brain endorphins, including beta endorphin. In the monkey brain the highest beta-endorphin levels are found in the interpenduncular, followed by the habenula and the hypothalamic subareas. High amounts of beta-endorphins are also detected in the preoptic areas, the substantia nigra, the pallidum and the superior and inferior colliculi. Low values are found in the limbic cortex and the cerebral cortical area while the cerebellar cortex contains the lowest value. In human plasma, the level of beta-endorphin is about 21+7.3 pg/ml or 6.2+2.2 f mole/ml.

Similarly, beta-LPH in the rat brain is detected in the hypothalamus, periventricular arcus, nucleus of the thalamus, ansa lenticularis, zona compacta of the substantia nigra, medial amygdaloid nucleus, zona incerta, PAG, locus coeruleus and a few fibres in the reticular formation. In human plasma, beta-LPH is from <20-150 pg/ml. Another peptide, des-Tyrl-gamma-endorphin, may be derived from beta-LPH, beta-endorphin or gamma-endorphin. It has been detected in the pituitary gland and brain. It has been shown that this peptide has no opioid activity, but can interact with neuroleptic binding sites in various areas of the rat brain.

(iii). Dynorphin

Goldstein and colleagues successfully identified a novel pituitary endorphin – dynorphin. It is a tridecapeptide containing [Leu]-enkephalin and is 700 times more potent than leu-enkephalin in inhibiting the guinea pig ileum longitudinal muscle contraction. This effect is only partially reversed by naloxone. Dynorphin is found in the posterior pituitary and spinal cord.

(iv). Other Related Substances

(a) **Endorphin Releasing factor** – A dipeptide, H-Tyr-Arg-OH, is found to be an endorphin releasing factor in the hypothalamus and has an analgesic potency 4.2 times higher than morphine.

(b) **Endogenous anti-opiate substance** - Cholecystokinin octapeptide appears to be the substance which is physiologically induced either by chronic morphine or chronic acupuncture treatment. Intraventricular injection of cholecystokinin octapeptide reverses morphine or acupuncture analgesia in naive animals.

4. Functions of Endorphins

In general, endogenous opioid peptides act similarly to morphine, which has a profound effect on the bodily functions (both physiological and behavioural) of humans and animals. These are different types of opiate receptors and endorphins which are located in numerous areas of the CNS and other parts of the body. The functions of endorphins are many and varied. The physiological and behavioural variability and characteristics of endorphin action depend on the location, action and types of endorphin-opiate receptor activities. In the past few years, met-enkephalins, leu-enkephalins, beta-enkephalins, dynorphin and their analogues, as well as other endorphins have been extensively studied. Their effects on humans and animals can be summarized in the following list:

(i). Analgesia - relieve pain and modulate stress

(ii). Homeostasis

(a) Releasing regulator of other hormones

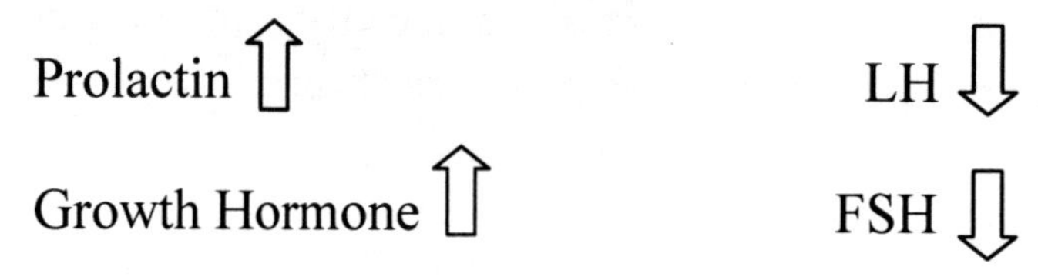

Thyrotrophin ⇧ Release oxytocin

Corticosterone synthesis ⇧ and vasopressin
(Beta-endorphin and ACTH secreted concomitantly)

(b) **Thermoregulation**
Hypothermia (Alpha-endorphin)
Hyperthermia (Beta-endorphin)

(c) **Digestive system** – Inhibit pancreatic secretions and increase insulin and glucagon release. Suppress intestinal motility in small intestine.

(d) **Cardiovascular system** – Depress heart and blood pressure.

(e) **Respiratory system** – Depress respiration.

(iii). Mental Illness

(a) Schizophrenia (abnormal high level of CSF endorphins, Leu(5)-beta-endorphin found in kidney dialysis).

(b) Small dose of beta-endrophin causes catatonia (akinesis), epilepsy and limbic seizures.

(c) High doses of naloxone reduces hallucination of schizophrenic patients.

(d) Des-Tyr-gamma-endorphin – a neuroleptic substance.

(e) Biopolar affective disorder:
Mania: high endorphin
Depression: low endorphin

(iv). Behaviour

(a) Interact with other neurotransmitters – acetylcholine, dopamine, GABA, 5HT, noradrenaline.

(b) Maintain normal behaviour –
alpha-endorphin (tranquilization)
gamma-endorphin (agitation and violent behaviour)
beta-endorphin (catatonia)
morphine and endorphin (euphoria)
leu-enkephalin (reward-pleasure-learning)
abnormally high pituitary beta-endorphin (obesity)
des-tyr-gamma-endorphin (tranquilizer)

(c) Sexual regulation –
High dose (6 μg) suppressed copulatory behaviour of rats.
Low dose (3 μg) increased mounting and intromission latencies.
Naloxone induces successful copulatory behaviour in sexual inactive rats.

(d) Naloxone was found to enhance memory and morphine antagonized this effect.
(e) Naloxone has also been used to treat hypovolemic shock; alcholic, diazepam, phenobarbital intoxication or withdrawal.

(v). Addiction: Morphine and endorphin are addictive.

(i) Analgesia

It has been verified that endorphins mimic the effect of opiates. Opiate receptors appear most densely in the paleospinothalamic pathway that transmits diffuse pain, and causes analgesia. Experimental evidence indicates that analgesia is observed after microinjection of met-enkephalin (120 μg/ injection) into or near ventral, caudal midbrain, and the PAG, while seizures and other pathological EEG changes are seen with injections into or near the forebrain dorsomedial nucleus of the thalamus. The noxious response of thalamic nociceptive neurons is also depressed by systemic or iontophoretic injections of D-ala(2) or D-Leu(5)-enkephalins. Met-enkephalin (0.2 – 20 μg / rat) and leu-enkephalin (1-20 μg / rat) produced a dose related and naloxone-antagonizable analgesia when microinjected into the nucleus reticularis gigantocellularis and nucleus reticularis paragigantocellularis of the medula oblongata. Microiontophoretical application of met-enkephalin into laminae I, II and III selectively suppresses the excitability of nociceptive dorsal horn neurons located in laminae V.

Enkephalin analgesia is one type of endogenous pain-control system which functions at the brainstem and spinal cord levels. Met-enkephalin acts like neurotransmitters that are released through the synapses of nerve terminals. Enkephalin-degradating enzymes (enkephalinase) and opiate receptors have markedly heterogeneous and parallel distribution amongst regions of the mouse brain. Extremely rapid degradation of met-enkephalin is observed in vivo and vitro. Thus an enkephalin pain-control system tends to be localized at the segmental or regional areas of the body and the analgesic effect is short lived.

The total brain enkephalin levels (by radio immunoassay, RIA) represent only 2-13% of the total endorphins radioreceptor assay (RRA). Perhaps the other endorphins may play a role for more profound, generalized and longer-lasting pain-relief (e.g. beta-endorphin and dynorphin)

Evidence indicates that intraventricular and intravenous administration of beta-endorphin can induce analgesia in human and animals. Stimulation-produced analgesia by implanting electrodes at the PAG or periventricular brain areas increases beta-endorphin in human ventricular CSF by 50-300% while stimulating the posterior limb of the internal capsule has no such effect. Foot-shock-induced analgesia is reversed by naloxone and is cross-tolerant with morphine in mice. This, however is in contrast with the results observed by Akil et al in rats in which stress-induced analgesia shows no cross-tolerance to morphine. This foot-shock-induced stress produces naloxone-reversible hyperthermia and increases plasma beta-endorphin levels six-fold. Beta-LPH, beta-MSH and beta-endorphin can pass from the circulation to the CSF and circulating beta-LPH can be cleaved to beta-endorphin. Beta-endorphin has a half-life of 9.3+0.75 minutes, but the pain-relief induced by systemic injection of beta-endorphin is long lasting and throughout the whole body.

Dynorphin is found in the spinal cord and posterior pituitary. It can produce a profound analgesic effect.

Naloxone reverses endotoxin hypotension and has great potential therapeutic value for shock (e.g., severe blood loss). While these results show that endorphins play a role in shock, the true mechanism has yet to be explored.

It was found that synthetic analogues of enkephalins, made by changing the amino acid sequence, can resist enzymatic degradation and can pass through the blood-brain barrier. Thus they can be taken orally to cause analgesia, and their analgesic potency can be tremendously increased (even up to 30,000 fold as compared to met-enkephalin). Unfortunately, they are highly addictive drugs.

(ii) Homeostasis

High concentrations of endorphins in the hypothalamus and adenohypophysis suggest that they play a role in pituitary endocrinological functions. Enkephalin neurons project from the hypothalamus to the pars nervosa in the pituitary and may be involved in the regulation of neurohypophyseal neurosecretion. Enkephalins and beta-endorphins have been shown to stimulate prolactin, growth hormone and thyrotrophin, but depress luteinizing hormone and thyroid stimulating hormone and FSH secretions. It has also been demonstrated that beta-endorphin releases oxytocin and D-ala(2)-enkephalinamide releases vasopressin. Beta-endorphin stimulates corticosterone synthesis in isolated rat adrenal cells by binding to the adenocorticotropic hormone receptors. ACTH and beta-endorphin are stored in the same secretory granules of the anterior pituitary, and are secreted concomitantly. There is a feedback loop between the beta endorphin (or ACTH) and corticosterone (or cortisol); while dehydration or dexamethasone administration reduces beta-endorphin and ACTH secretion, adrenalectomy or metyrapone administration elevates plasma beta-endorphin levels sevenfold. Endorphins in the pituitary seem to be releasing factors or precursors to releasing factors of corticosterone. Futhermore, in normal animals, naloxone alone produced opposite effects in modulating the hormonal release. This suggests that endorphins participate in regulating normal secretions of the pituitary. Enkephalinergic cells are found in the islets of Langerhans and beta-endorphin and enkephalins have been shown to inhibit pancreatic secretions and increase insulin and glucagon release. Large amounts of opiate receptors are found in the intestine and stimulation of enkephalinergic neurons in the small intestine suppresses intestinal motility. Beta-endorphin produces hypothermia and alpha-endorphin produces hyperthermia. Finally, met-enkephalin has been shown to depress heart rate, blood pressure and respiration when applied to the brain stem. All the above results indicate that endorphins are involved in the physiological regulation of the cardiovascular, respiratory, and digestive systems and body temperature.

(iii) Mental illness

Endorphins have been shown to produce profound physiological effects when administered directly into the brain. They cause “wet-dog” shakes, sedation, catatonia and catalepsy. Seizures and long lasting epileptiform changes in EEG are observed with injections of met-enkephalin (170 μg/injection) into or near the forebrain dorsomedial nucleus of the thalamus or leu-enkephalin (1-20 μg) into the dorsal hippocampus. Low (non-analgesic) doses of beta-endorphin induce non-conculsive limbic seizures and catalepsy.

Beta-endorphin has been used to induce akinesia in animals and a significant increase in concentration of 5-hydroxytryptamine is observed in the midbrain. The induced akinesia can be reversed by methylmorphine, L-dopa and partially by L-amphetamines. It has been demonstrated that beta-endorphin or d-ala(2)-enkephalin analogue increase serotonin metabolism in the hypothalamus and dopamine turnover in neostriatum, tuberculum olfactorium and nucleus accumbens. It is proposed that there is a feedback relationship between endorphin neurons and dopaminergic neurons.

The potent and divergent behavioural responses of animals to natural occurring substances (especially beta-endorphin which has a more marked, prolonged effect, even at very minute doses) indicate that alteration in their homeostatic levels might have etiological significance in mental illness. CSF endorphin levels were found to be higher than normal in depressed or schizophrenic patients and lower than normal in chronic pain patients. This might explain how electric shock therapy has a therapeutic effect on schizophrenic patients as it might reduce their endorphin levels. Chronic schizophrenic patients under neuroleptic medication show rapid inactivation of CSF endorphin. Kappa and delta receptors may be the ones that mediate sedation, dysphoric and hallucinatory syndromes. High doses of naloxone have been reported to eliminate hallucinations in schizophrenic patients. Leu(5)-beta-endorphin is detected only in kidney dialysis of schizophrenic patients and it has been suggested that this may play a role in the chemical etiology of schizophrenia.

Van Ree and colleagues reported that des-Tyr(1)-gamma-endorphin is a neuroleptic substance; it demonstrates no opioid activity, but can interact with neuroleptic binding sites in various areas of the rat brain. Beta-endorphin and other opiates have no affinity for neuroleptic binding sites. Des-Tyr(1)-gamma-endorphin is found to be very effective in treating schizophrenic patients. They postulate a duo system whereby endorphin reduces stress and des-Tyr(1)-gamma-endorphin produces sedation.

(iv) Behaviour

Endorphinergic neurons are shown to have extensive interactions with acetylcholine, dopamine, 5HT, noradrenaline and GABA neuronal systems in the brain. Beta-endorphin increases the turnover rate of GABA in the substantia nigra and lobus, but decreases the turnover rate in the nucleus caudatus. The turnover rate of acetylcholine is decreased by beta-endorphin in the cortex, hippocampus nucleus accumbens and globus pallidus, but not in the nucleus caudatus. Met-enkephalin and leu-enkephalin also inhibit the spontaneous release of acetylcholine from the cerebral cortex in vivo. This indicates that endorphins may act as presynaptic inhibitors of cholinergic neurons. The interactions of endorphins with other neurosubstances may preserve a homeostatic state of bodily functioning which helps to maintain normal behaviour. Evidence has shown that alpha-endorphin induces tranquilization, gamma-endorphin produces agitation and sometimes violent behaviour, beta-endorphin causes catatonia and hypotension which is reduced by PCPA, and des-Tyr(1)-beta-endorphin improves schizophrenia. All these effects suggest that the endorphinergic systems are probably involved in maintaining normal behaviour and that a malfunction of this system may result in psychiatric illness or abnormal behaviour. For example, evidence indicates that overeating is probably caused by the abnormal high pituitary beta-endorphin level (5 times higher than normal animals) in obese mice (ob/ob) and rats (fa/fa). Naloxone selectively abolishes overeating in these genetically obese mice and rats.

Endorphins, like morphine, produce euphoria. It has been found that leu-enkephalin induces higher rates of self-administration in the self-administered bar-pressing behaviour as compared to met-enkephalin when injected into the limbic areas in rats. It is possible that leu-enkephalin is a major intrinsic substance that elicits the pleasure and reward system.

Social and sexual behaviours are also affected by endorphins. Beta-endorphins increase grooming activity, decrease mounting and increase the length of time before ejaculation in rats. A high dose (6 μg) of D-Ala(2)-met-enkephalinamide (analogue) also suppresses the copulatory behaviour of rats, but a low dose (3 μg) increases mounting and intromission latencies.

(v) Addiction

Endorphins, like morphine, produce tolerance, cross-tolerance and withdrawal effects on long-term treatment, i.e., they are also addictive. Endorphins can modify the relative activities of adenylate cyclase and guanylate cyclase, two enzymes floating freely in the double layer of lipid molecules that make up the cell membrane. These two enzymes will be activated to synthesize cyclic AMP and cyclic GMP when appropriate neurotransmitters bind to the membrane receptors. Often, these two enzymes display antagonistic action in mediating the same hormonal effect inside the cell. In neuroblastoma-glioma hybrid cell cultures, long term exposure to opiates or endorphins will induce an abnormal production of adenylate cyclase (which enhances cyclic AMP synthesis) to compensate for the opiate-inhibited level. As a result, more opiates or endorphins are required to produce decreases in cyclic AMP levels. On withdrawal of opiate or endorphin, cyclic AMP levels increase markedly owing to the abnormally high quantity of adenylate cyclase. This increase in cyclic AMP level correlates biochemically to withdrawal symptoms. The increase in the cyclases during and after long-term opiate or endorphin treatment may act as a biofeedback mechanism which will change the firing rate of endorphin neurons, including the development of tolerance and physical dependence. It has been shown that SPA of the PAG or chronic beta-endorphin administration cause tolerance which can be reversed by 5-

hydroxytryptophan. Thus serotonin may play an important role in addiction and analgesia.

Prolonged activation of opiate receptors by exogenous opioids causes increase in protein kinase C, which interferes with action of opiate receptor and increase effect of N-methyl-D-aspartate(NMDA) receptor activation, resulting in tolerance and hyperalgesia.

5. Enkephalinase

Met-enkephalin and leu-enkephalin are sensitive to trispin and chromotrispin, carboxypeptidase A and leu-aminopeptidase. It is suggested that rapid inactivation of enkephalins may be due to at least two different enzymes localized in different structures of the cell. A peptidyl dipeptidase, which is a component of plasma membrane, may degrade enkephalins by liberating a C-terminal dipeptide, while an aminopeptidase in endothelial cells may cleave peptides that are taken up as blood flows past the endothelial surface. Aminopeptidases are also found in the brain homogenates. Recently a high affinity enkephalin-dipeptidyl-carboxypeptidase (enkephalinase) was found to have markedly heterogenous and parallel distribution to opiate receptors in different regions of the mouse brain. The concentration of enkephalinase in areas of the mouse brain is in the order of:

Striatum>Hypothalamus>Cortex-Brainstem>Hippocampus>Cerebellum

Hippocampus-Cerebellum

Enkephalinase in the brain is increased after morphine administration, and kainic acid or 6-hydroxydopamine lesions of the nigrostriatal dopaminergic pathways lead to similar decreases in this peptidase and opiate receptors.

Peptidase-inhibitors such as puromycin inhibit the degradation of leu-enkephalin by enzymatic activities present in rat brain homogenates and guinea pig ileum, while bacitracin potentiated and prolonged the in vivo analgesic activities of beta-endorphin and acupuncture and analgesia. Inhibitors of carboxypeptidase A and leu-aminopeptidase, D-phenylalanine and D-leucine (respectively), also induce naloxone reversible analgesia in humans and mice. Thus drugs that protect endogenous endorphins from enzymatic degradation may be very useful for clinic pain relief. Recent biochemical evidence supports this hypothesis; D-phenylalanine and D-leucine can cross the blood-brain barrier and inhibit enkephalinases in guinea pig ileum assay. Combining the D-amino acids and EA treatments produced a higher analgesia in larger numbers of mice as compared to either treatment alone. It is postulated that EA releases endorphins which are protected by the D-amino acids, thus allowing production of a higher analgesia.

III. HISTORY OF ACUPUNCTURE IN CHINESE MEDICINE

The ancient art of treating diseases and relieving pain by acupuncture is an integral part of Chinese medicine. Although there is no longer a record of its initial discovery, some famous legends concerning acupuncture have persisted for about 5,000 years. According to one story, acupuncture was first discovered by a group of warriors, who after being wounded by arrows sometimes noted a sensation and "miraculously" recovered from ailments by which they had been plagued for many years (The Academy of Traditional Chinese Medicine, 1975). Another story tells that a wise man accidentally struck his lower leg against a sharp stone and relieved the pain in certain parts of his body. Accordingly, the first needles were made of sharp pieces of stones called "pien", and this work eventually came to refer the curing of diseases by pricking with a stone (*Shuo Wen Ji Zi Analytical Dictionary of Characters,* compiled during Han Dynasty (206 BC – A.D. 200). Later needles were made from bamboo, copper, iron, gold, silver and finally stainless steel. The steel needles frequently used at present vary in length from 1-20 cm and in thickness from 26-32 gauge (0.45- 0.26 mm in diameter).

The earliest record of a successful cure by means of acupuncture is described in a book called *Shik Chi (Historical Records)* written 2,000 years ago. The book states that Pien Chueh, a famous physician of Warring States Period (475 – 221 BC) used acupuncture to revive a dying patient already in a coma. The classical method of acupuncture, however, is first described in Ling Shu, a special chapter in *Wang Di Nei Ching (The Yellow Emporers's Book of Internal Medicine)* in which treatment of sickness, meridians and points, alleviation of pains in the head, ear, tooth, back, stomach, abdomen and the joints are discussed in detail and embellished with Chinese medical theories.

Tsin Dynasty (265 – 420 A.D.) Acupuncture was in popular usage and a more comprehensible and systematic text-book named the "*Chen Chiu Chia Yi Ching (An Introduction to Acupuncture and Moxibustion)* was published. This book listed 649 acupuncture points, mapped out 349 basic points on the human body and clearly reviewed the theory of acupuncture and the needling techniques.

Tang Dynasty (618 – 907 A.D.) Acupuncture was taught in the Imperial Medical College organized by the Chinese government, and the method of needling spread to Korea, Japan and India.

Ming Dynasty (1364 – 1644 A.D.) Extensive records of practical knowledge were accumulated and summarized in a book known as the "*Chen Chiu Ta Cheng*" *(Compendium of Acupuncture and Moxibustion)* which is still in common use today.

During the **Ching Dynasty** (1644 – 1911) and Nationalist Chinese rule (1911 – 1949), the practice of acupuncture was suppressed, and the authorities banned the art. During this time China was being exposed to the vast body of Western science and medicine, yet the practice of acupuncture and other traditional healing arts still persisted among the common people. Often knowledge was kept secret within families and passed along from father to son. After the founding of the People's Republic of China, an attempt was made to combine the ancient traditional healing arts with the relatively newly acquired Western technology.

There were many new developments during this last time period, such as the use of electronic instruments and acupuncture anaesthesia starting in 1958. Since then, more than 400,000 major and minor surgical operations have been performed on both adults and children by means of acupuncture anaesthesia.

Acupuncture was first introduced in Europe by a German, Dr. E. Kampfer in 1683. The significant evaluation of this subject, however, was done by George Soulie' de Morant who wrote: *L'Acupuncture en Chine et la Reflextherapie Moderne* and *Les Arguilles et les Moxas en Chine* in 1863. Gradually, acupuncture was accepted by some countries in Europe, notably France, Germany and Austria. Only very recently (after President Richard Nixon's 1970 visit to China), was acupuncture introduced to the West, specifically North America. A surge of interest and scientific investigations in the ancient oriental art began to emerge.

IV. THE CLASSICAL THEORY OF ACUPUNCTURE

The classical theories of Chinese medicine are integrated with the ancient philosophy called Taoism which describes the universe as being composed of a cosmic force field in which two basic elements, Yin and Yang, are perpetual complements in continuous change. Yang, the positive, corresponds to such things as sun, day, heat, light, dryness, male and life. Ying, the negative, corresponds to the moon, night cold, darkness, water, female, death and many others. Yin and Yang are said to be dynamically opposed, yet are harmonizing energies in the universe. Accordingly, these two elements are considered to comprise the balance of man's 'life energy' called "chi", and the human body is treated as a small universe system. The chi circulates continuously throughout the body along invisible pathways known as meridians, each of which originates in one of the principal internal organs and then surfaces to run along the outside the body, sometimes as close as a millimetre or two beneath the skin. The meridians are able to transmit signals of internal illness in any major organ to the outside, and can also transmit stimuli back to internal organs from the body surface.

In all, there are seventy-one meridians in the human body, classified into channels (Ching) and collaterals (Luo'). Channels are the main pathways running lengthwise and are made up of twelve main channels and eight extra-channels, while the collaterals are divided into major collaterals and sub-collaterals which connect one channel to another. Hence the entire 'Chingluo' system is distributed over the whole body and connects the viscera with the four extremities, skin and the sense organs, making the body an organic whole.

The twelve Channels are:

1. Lung Channel of Hand-Taiyin (Lu).

2. Large Intestine Channel of Hand-Yangming (L.I.).

3. Stomach Channel of Foot-Yangming (St.).

4. Spleen Channel of Foot Taiying (Sp.).

5. Heart Channel of Hand-Shaoyin (H.).

6. Small intestine Channel of Hand-Taiyang (S.I.).

7. Urinary Bladder Channel of Foot-Taiyang (U.B.).

8. Kidney Channel of Foot-Shaoyin (ki).

9. Pericardium Channel of Hand Jueyin (P.).

10. Triple Warmer of Hand-shaoyang (T.W.).

11. Gall Bladder Channel of Foot-Shaoyang (G.B.).

12. Liver Channel of Foot-Jueyin (Liv.)

There are also two important meridians running along the midline of the body:

1. Governing vessel (Gv) runs along the back of the body midline.

2. Conception vessel (Cv) runs along the front of the body midline. (see Appendix I)

More than 360 acupuncture points lie along all these meridians and can be used to affect the internal organs of the body for treating illnesses. Traditional theory states that sickness arises when there is too much Yang (over-tonification) or too much Yin (over-sedation). Therefore the two forces are not in balance and the chi does not circulate smoothly throughout the meridians. Acupuncture will either tonify or sedate the body to restore the smooth flow of this 'life energy', hence balancing the Yang and Yin forces and eliminating the illness.

It should be borne in mind that this is not a 'scientific theory', but rather a methodology based on an ancient philosophical system. It does, however, reveal the profound difference between the systems of thought in the East and the West.

V. New Meridian Theory

In ancient times, medical workers did not understand how acupuncture worked. They used philosophy to make up the classical theory of Chinese meridians to explain the observations and experiences of acupuncture effects. Through scientific studies, the clinical responses of acupuncture can be achieved by stimulating the peripheral nervous systems and/or the autonomic nervous system. Almost all the acupuncture points lie on the peripheral nerves or autonomic nerves. All the autonomic nerves (either the sympathetic or parasympathetic nerves) run parallel to the venous and arterial systems. To achieve acupuncture effects, it must stimulate the peripheral nerves (sensory or motor nerves) and/or the autonomic nerves. Experiments have demonstrated that there are no acupuncture effects at all if these nerves are blocked by xylocaine. By blocking the nerves, there is no endorphins-release for analgesia nor cortisol-increase for anti-inflammation. Therefore, the Chinese meridian theory is merely related to the peripheral nervous system and the autonomic (parasympathetic and sympathetic) nervous system.

For interest and comparison, some ideas of ancient classical acupuncture theory are listed below and the recent scientific discovery and theory of acupuncture are also presented in detail:

Classical Theory of Acupuncture
Figure 1. The Circadian Rhythm of Meridians

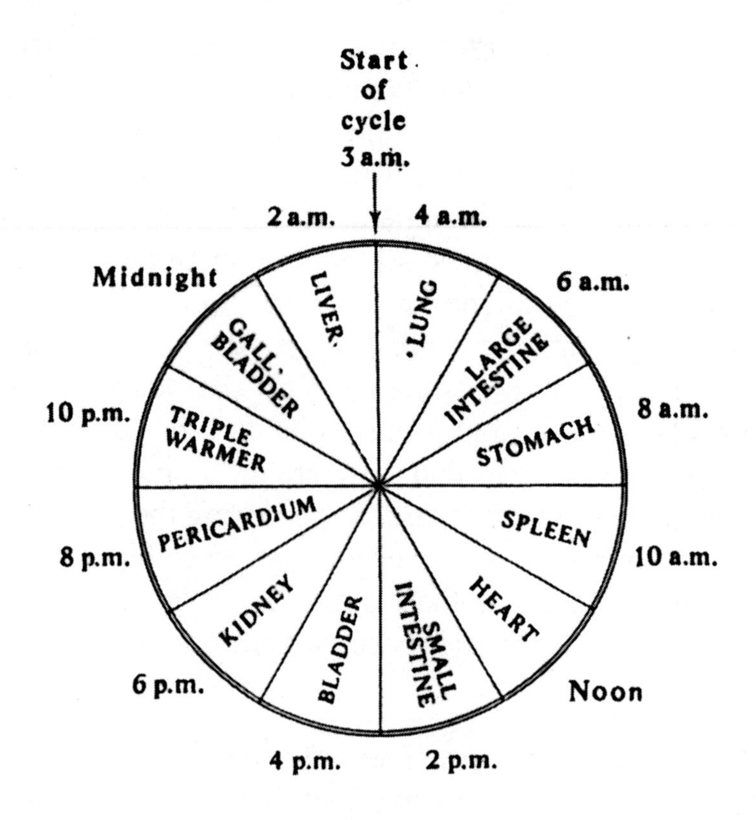

Classical Theory of Acupuncture
Figure 2. The Relationship of the Five Elements

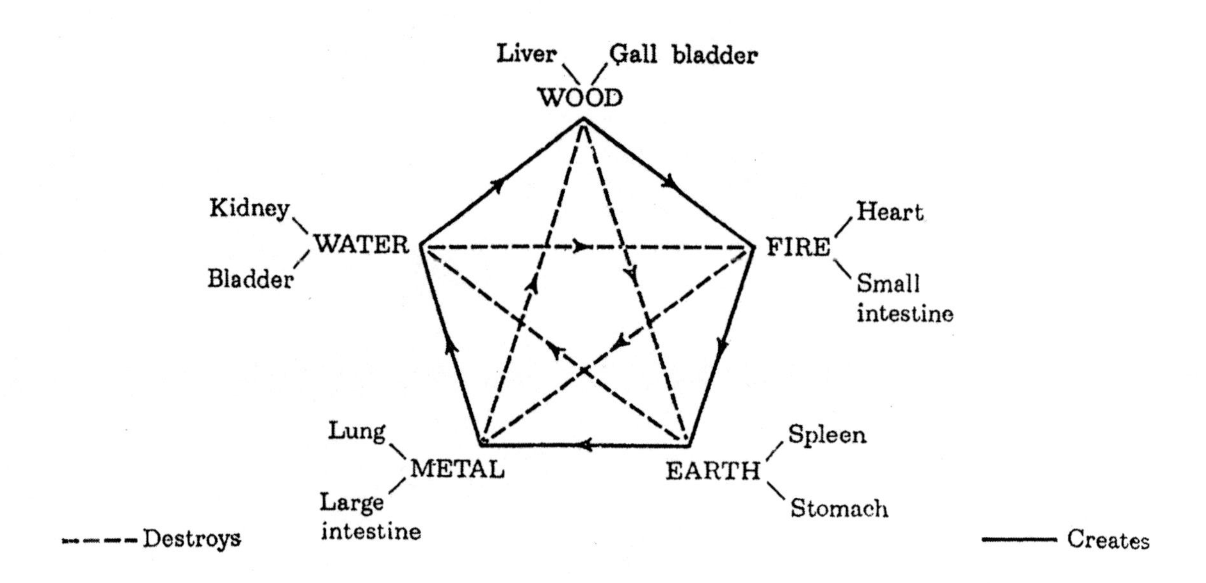

Classical Theory of Acupuncture

Figure 3a. The Relationship between Meridians and the Five Elements

In the hands, three yang meridians begin at the fingertips on the dorsal side; and three yin meridians end at ventral fingertips.

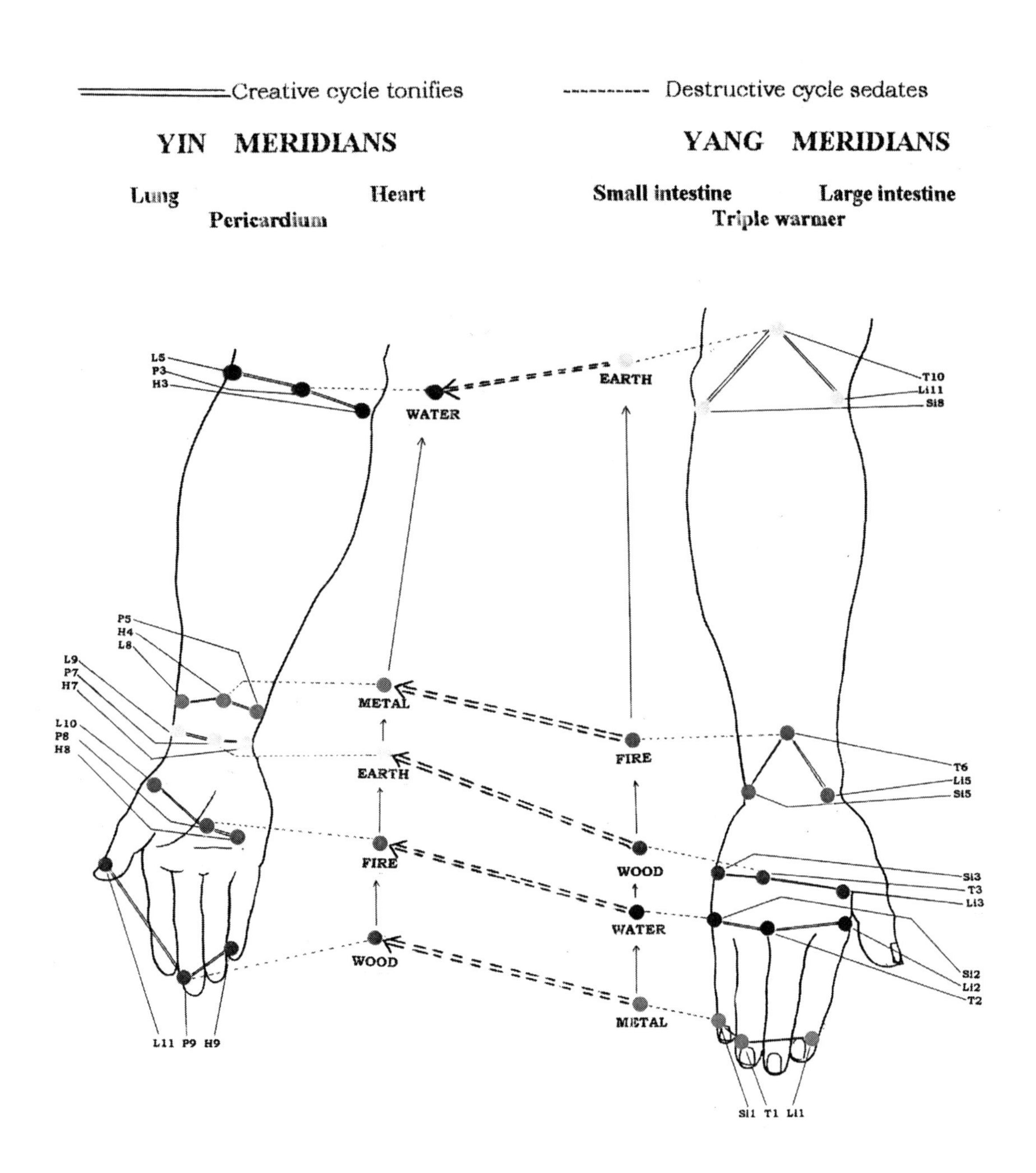

Classical Theory of Acupuncture

Figure 3b. The Relationship between Meridians and the Five Elements

In the feet, there are three Yin meridians that begin from the medial side of the feet and legs, and three Yang meridians that end at the lateral side of the feet and legs.

═══════ Creative cycle tonifies ---------- Destructive cycle sedates

YIN MERIDIANS — Spleen, Liver, Kidney

YANG MERIDIANS — Stomach, Gall, Bladder

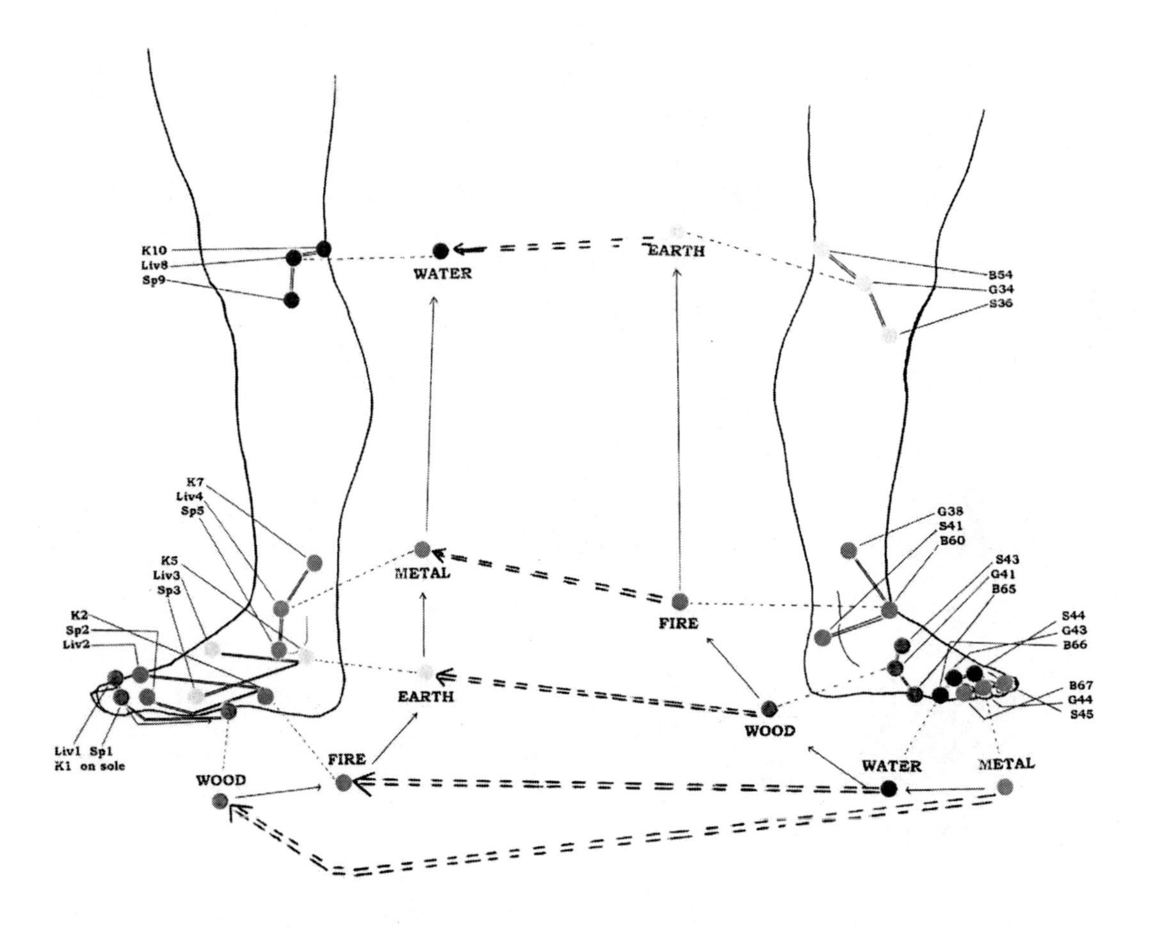

Classical Theory of Acupuncture
Figure 4. Points of Tonification (to tonify the corresponding meridians)

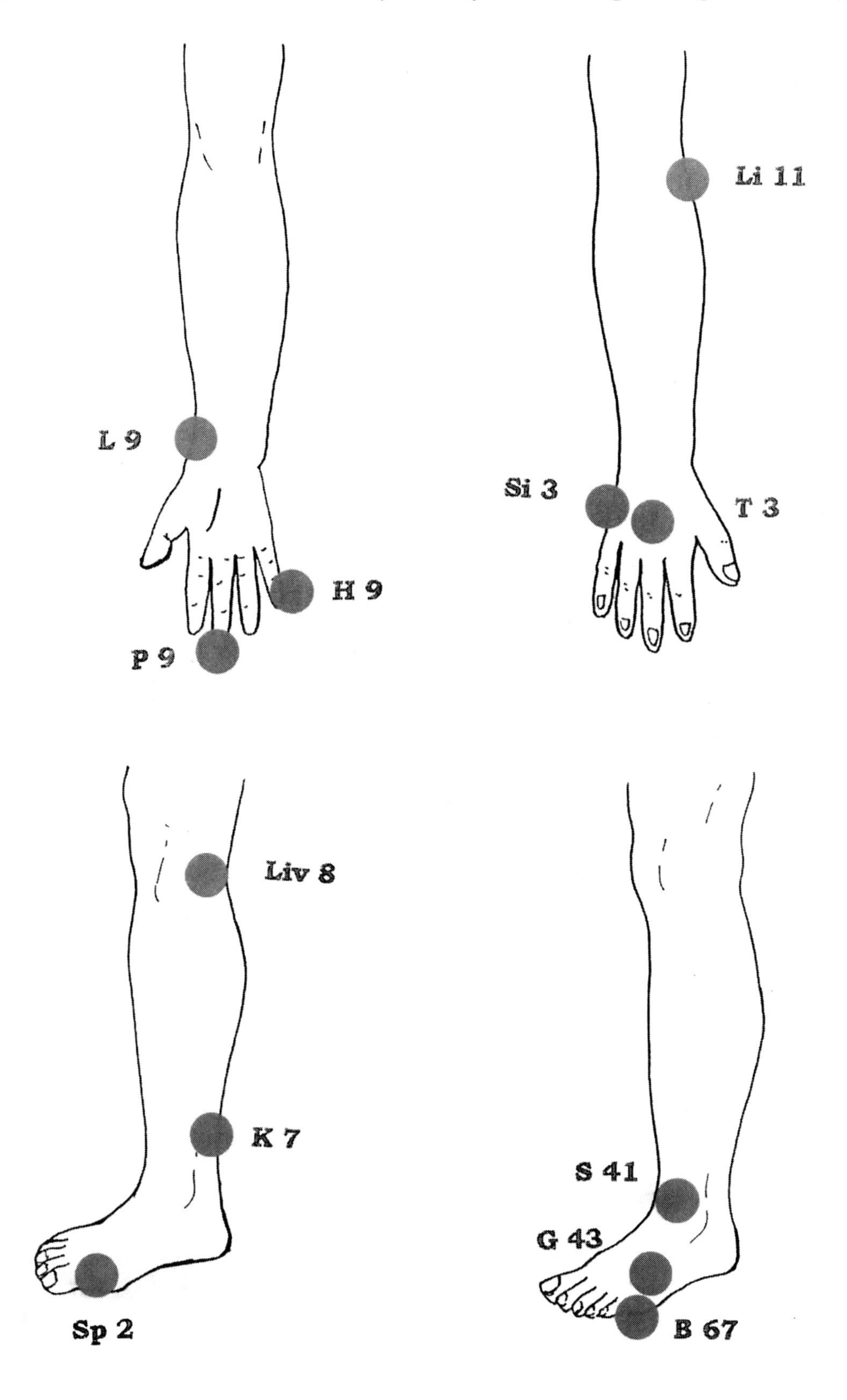

Classical Theory of Acupuncture
Figure 5. Points of Sedation (to sedate the corresponding meridians)

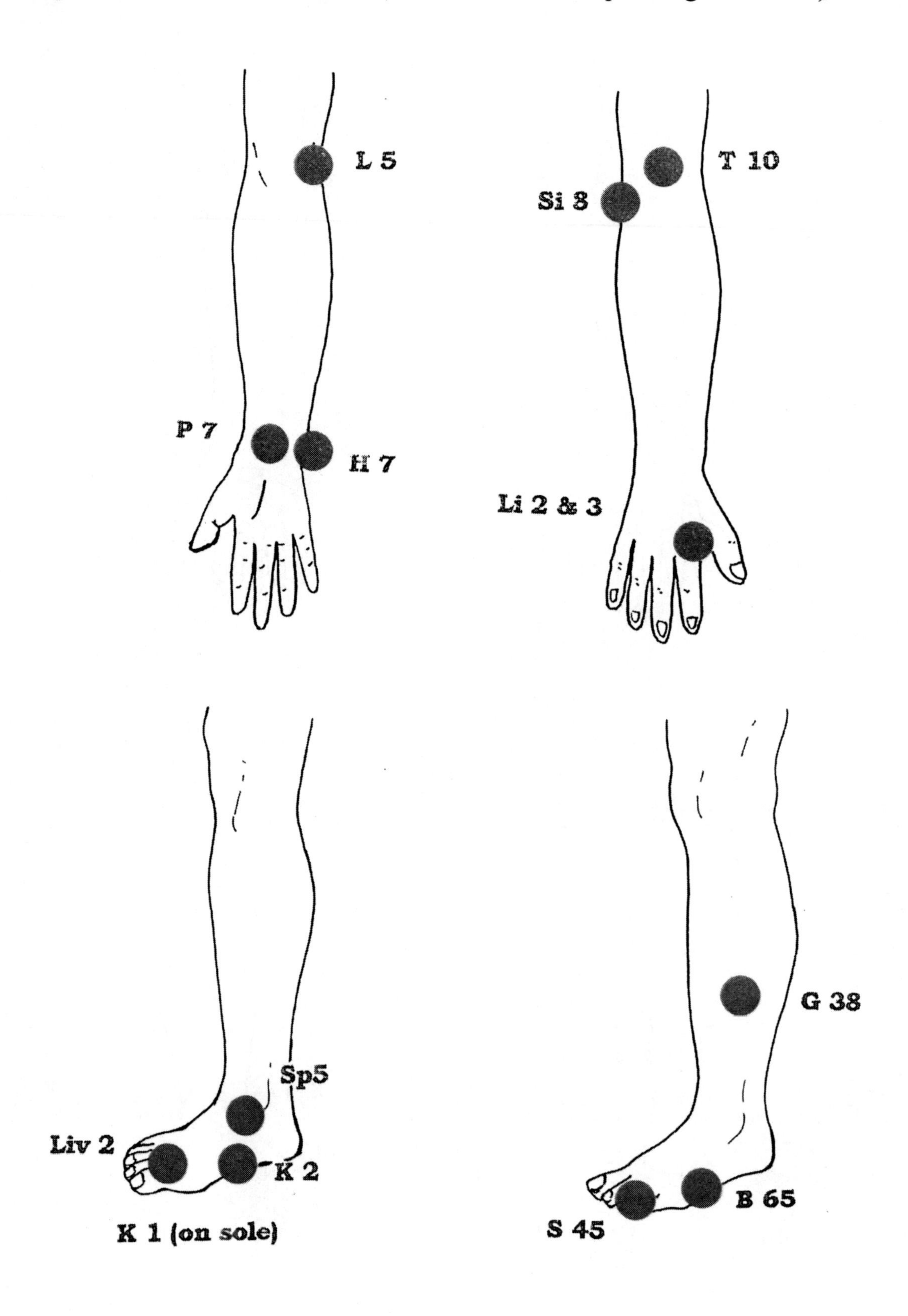

For tonification, a gold needle is inserted in the direction of the energy flow of the meridians, then rotated clockwise while the patient exhales. For sedation, a silver needle is inserted in the direction of the energy flow of the meridians, and rotated counterclockwise while the patient inhales. In clinical practice, tonification and sedation have similar acupuncture effects. Only stronger mechanical or electrical stimulation has a stronger analgesic effect.

Classical Theory of Acupuncture
Figure 6. The Connecting Points

Two meridians can be stimulated by one acupuncture point. For example, stimulating G37 can balance the energy flow in both the gallbladder and liver meridians.

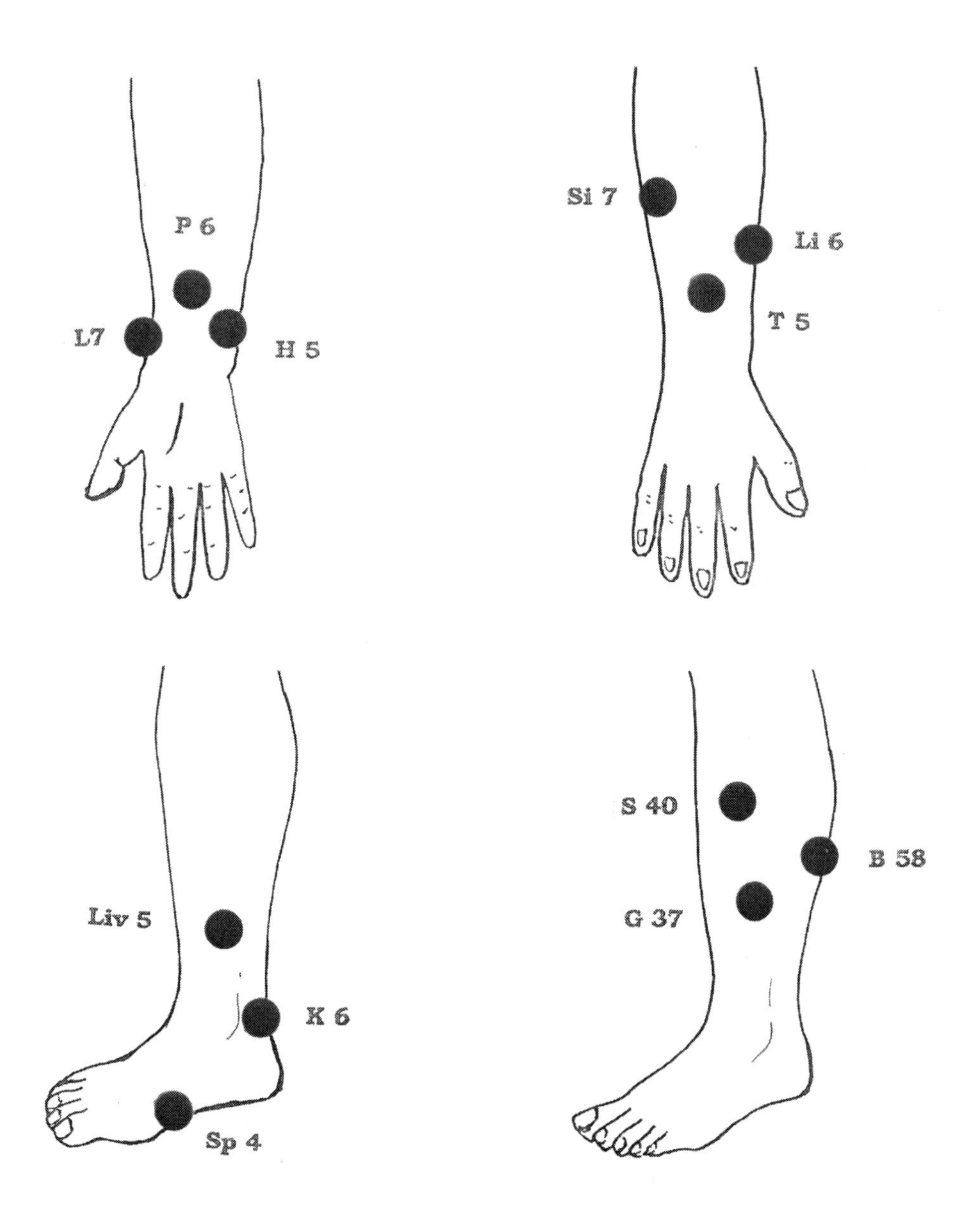

Classical Theory of Acupuncture

Figure 7. The Source Points of the Meridians

The source points are often more effective in treating the disease that the meridian represents, e.g., He-7 for heart meridian.

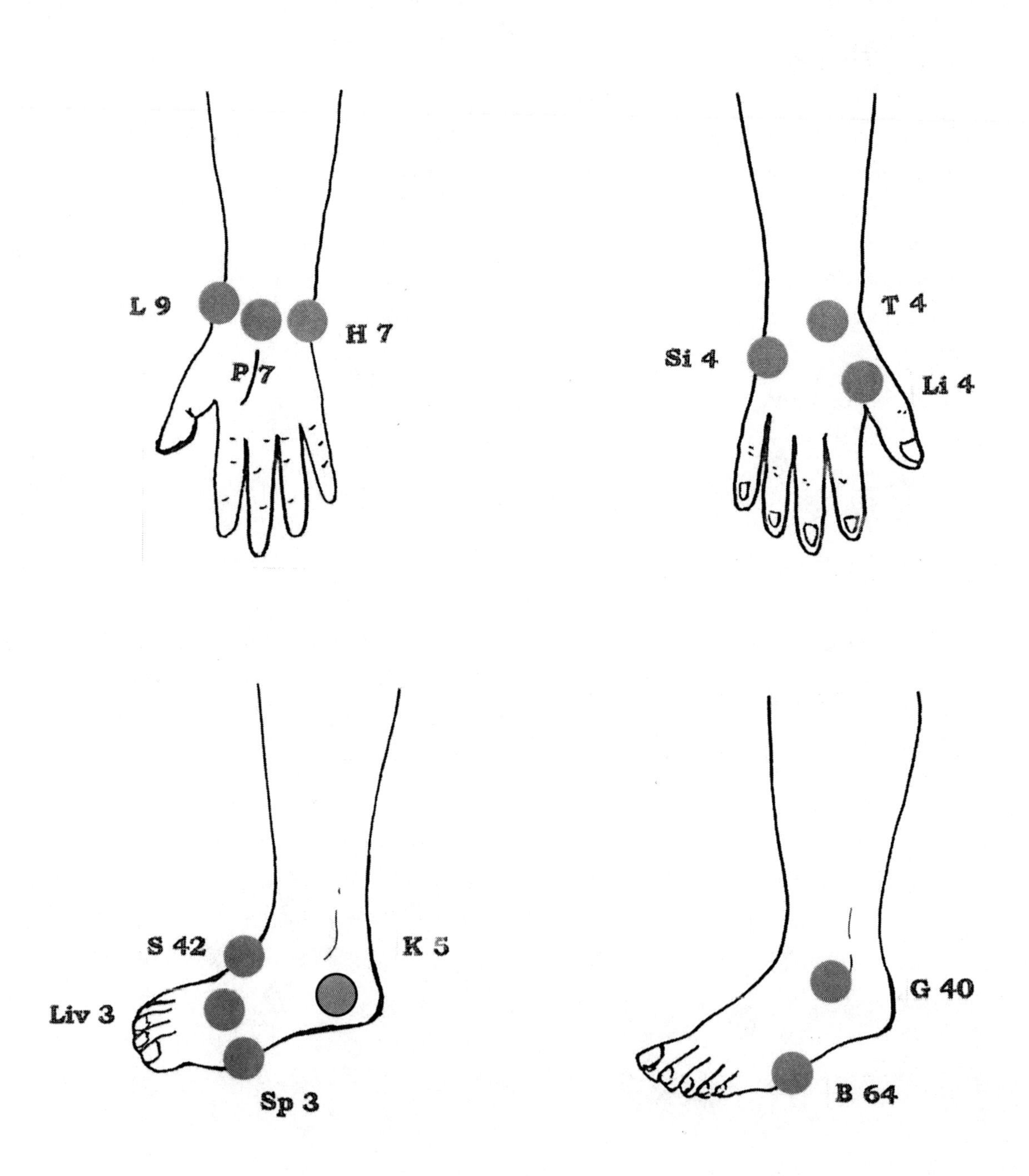

Classical Theory of Acupuncture
Figure 8. Pulse Diagnosis

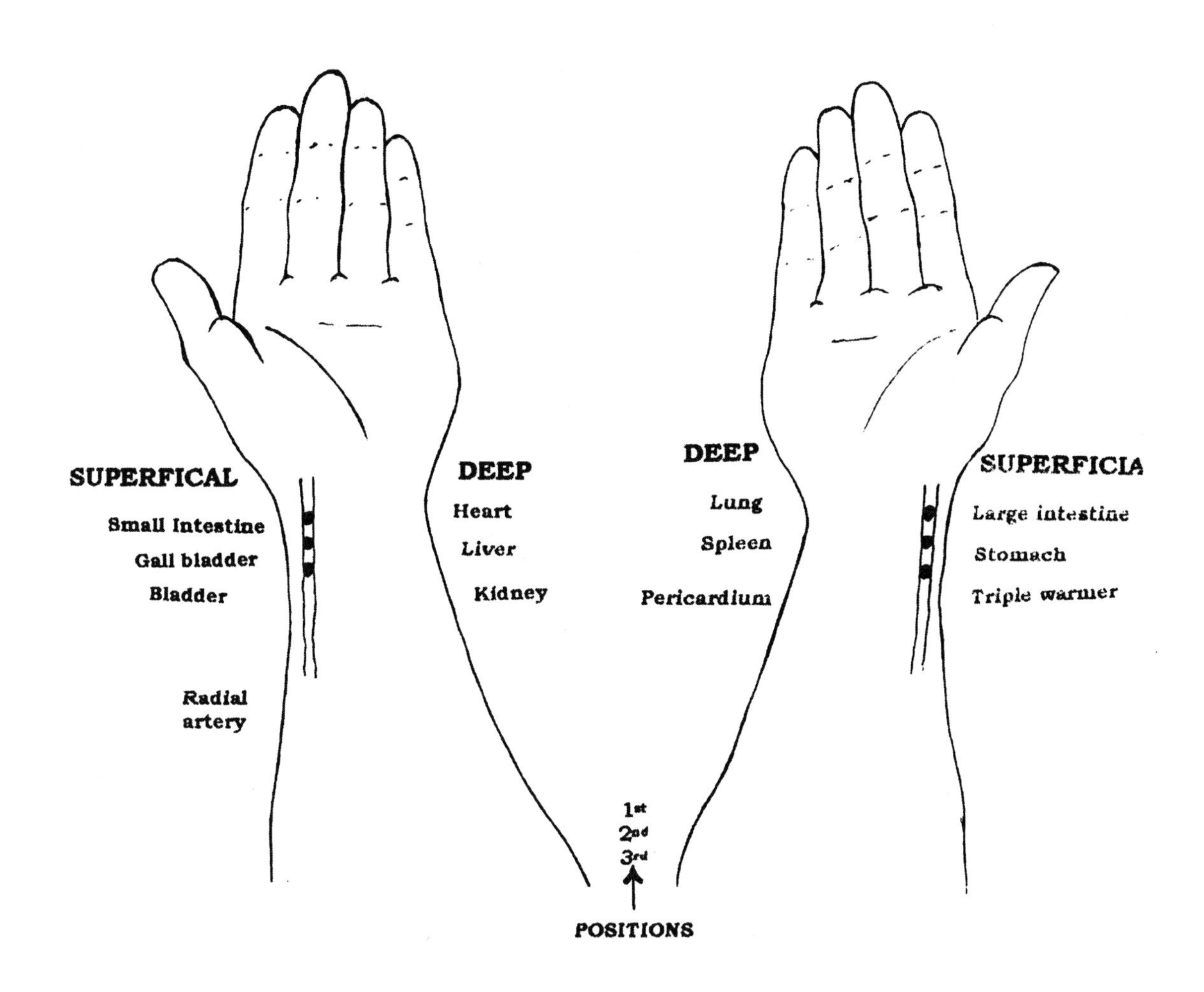

In classical Chinese medicine, it is necessary to feel the radial artery pulses before a diagnostic judgement is made to either sedate or tonify the diseased meridians. As shown in Figure 8, the pulse at the radial artery of the wrist is divided into three zones, each of which has a superficial and deep position,.

Pulse diagnosis may take years to learn, yet is neither reliable nor consistent (five acupuncturists may have five different interpretations). More an art than a science, it has many limitations and uncertainties compared to modern medical diagnosis.

Western physicians may find these ideas interesting, fascinating, profound and mysterious. Many traditional Chinese medical workers firmly believe in this phenomenon of five elements, yin and yang, sedation and tonification, etc. If you can follow these laws and instructions in your practice of acupuncture, good results may be achieved most of the time. This methodology is based on an ancient system of philosophy, and often depends on pulse diagnosis. It is not easy to learn, understand and follow the exact pattern of the laws. Different healers may have different skills and interpretations of these pulse diagnosis, the laws of five elements, mother and son, husband and wife, yin and yang, 24 hours and seasonal cycle of meridians, etc. We have to remember that this is not a scientific theory, but rather our ancestors using an ancient system of philosophy to explain the observed methodology which they did not understand.

Even if we do not follow the above laws of five elements, sedation and tonification, etc., we may often find that acupuncture, EA and TENS SUPERFICIAL (Transcutaneous Electrical Nerve Stimulator) can give a similar or even better result when a modern diagnosis is applied (including medical history, physical examination and scientific investigations). Tonification or sedation of the acupuncture point makes no difference in treating an illness. The only difference is that some patients need stronger, intense stimulation and some need weaker, less intense stimulation to get a better effect. The most effective five element acupuncture points and the source points for the meridians lie between the fingertips, the elbow, between the tips of the toes and the knees, a series of points whereby it is possible to treat nearly any disease. From an anatomical point of view, it is easier to stimulate the peripheral and autonomic nerves in the hands and feet ,which have a more intense and larger territorial sensory input into the brain. This will ensure the release of neurochemicals and hormones for pain relief and disease treatments as explained in VI Mechanism of acupuncture, EA and TENS.

VI. HISTORY OF ELECTROTHERAPY[5,6,7]

The use of electrical stimulation for medical purposes was described in the books of Hippocrates in about 420 BC. The Roman physician Scribnius Larguz used the torpedo fish or electric ray to treat pain, especially headache and gout. This method of bioelectric therapy was also used to treat hemorrhoids by a Greek physician, Diososides. The torpedo fish continued to be used until the 16th century to treat migraine, melancholy and epilepsy. In 1672, Otto von Guericke constructed an early prototype of an electrostatic generator. In Germany in 1744, Christian Gottlieb Kratzenstein was able to use a modified electrostatic generator to treat paralysis, epilepsy, kidney stones, sciatica, and angina pectoris. In the mid-19th century, G.B. Duchenne, often called the father of electrotherapy, identified the motor points and the action of all muscles in the body and proposed the use of Faradic (induced) current and moistened pads as surface-electrodes for electrical stimulation. By 1900, most American doctors had at least one electrical machine in their offices to treat painful symptoms of rheumatism, gout, neuralgia, bruises, fracture and even insomnia. In 1960, it was found that connecting the acupuncture needles to an electrical stimulator produced stronger analgesic effects and in the 1970s electroacupuncture (EA) followed by Transcutaneous Electrical Nerve Stimulation (TENS) enjoyed widespread use.[1] In the decade that followed, Dr. C.N. Shealy, an American neurosurgeon, was the first to implant electrodes into the spinal cord and brain for the purpose of producing analgesia by direct electrical stimulation.[2] This led to the engineering of the first modern TENS machines for the treatment of pain in North America in 1973.[3]

VII. THE MECHANISMS OF TENS AND ACUPUNCTURE

Abstract

Acupuncture, electroacupuncture (EA) and TENS are widely used by many medical workers especially in physiotherapy, pain and sport medicine clinics. Many researchers have tried to look for the plausible and interesting mechanisms, particularly in its efficacy in the treatment of many pain problems. Different levels of analgesia and clinical benefit can be achieved by changing treatment parameters, such as current frequencies, intensities and electrode placements, etc. Possible anatomical sites and mechanisms of TENS and EA as related to endorphins and monamines are proposed.

The Use of TENS and Acupuncture

In China, acupuncture has been used for several thousand years, not only to alleviate pain, but also to treat depression, tinnitus, insomnia, hay fever, asthma, psoriasis, dermatitis, the common cold, irritable bowel syndrome, cigarette withdrawal symptoms, drug or alcohol addiction, and musculoskeletal and neurovascular problems, acute low back and neck spasms and other myofascial pain, herniated intervertebral disks of the spine and degenerative back-pain. Other treatable conditions are migraine headaches, tension headaches, carpal tunnel syndrome, sinusitis, intercostal neuralgia, post-herpetic neuralgia, trigeminal neuralgia, diabetic neuropathy, leg growing pain, etc. Most pain problems that generally respond well to acupuncture and TENS are musculoskeletal in origin, including osteo- and inflammatory arthritis, tendonitis/ bursitis of the shoulders, elbows, wrists, hands, hips, knees, ankle and foot, work- and sports-related strains and sprains.

The Mechanisms of TENS and Acupuncture
Table 1. A total number of 362 patients were treated in the Sports Medicine & Pain Management Clinic during a typical week

Table 1. Pain clinic population by anatomical distribution of presenting complaint.

Age	Low back	Neck	Shoulder	Upper extrem.	Lower extrem.	Head	Other	Sample size
60+	31	16	9	3	21	1	4	85
45-59	55	28	21	22	35	7	7	175
25-39	30	13	4	10	13	4	6	80
<25	6	3	1	2	2	2	6	22
Totals	**122**	**60**	**35**	**37**	**71**	**14**	**23**	**362**

The analgesic effect of TENS or electroacupuncture ranges from 30-90% in most patients.[4] Certain common problems such as soft tissue injuries (tendonitis, sprain and strain), osteoarthritis, tension headache, etc., often respond very well to electrical stimulation and will recover quickly and completely. Synergistic results, however can be achieved by combining a variety of other therapeutic approaches such as medications (NSAIDs, analgesics, anticonvulsants, antidepressants, SSRJs, antimigraine), trigger point injections, strength and aerobic exercises, spinal traction, manipulation, traditional Chinese herbs, etc. While TENS and electroacupuncture can be very useful to help many diseases, they should not be the sole method of treatment for illnesses like hypertension, anorexia, colitis, enterocolitis, hepatitis, nephritis, prostatitis, prostatic hypertrophy, Alzheimer's disease, Meniere's disease, severe asthma, depression, rheumatoid arthritis, lupus, migraine headache, chronic back pain, frozen shoulders, diabetic neuropathy, tinnitus, fibroid, menopause syndrome, etc.

Much research shows that acupuncture released endogenous opioid substances in the spinal cord and brain of animal models and humans.[8,9] In 1979, while completing doctoral work at the University of Toronto in the laboratories of Dr. B. Pomeranz,, the author (Dr. R. Cheng) first proposed that different frequencies of electrical stimulation could trigger the preferential activation of different endogenous pain-relieving pathways.[10] It was suggested that stimulation at 4 Hz released beta-endorphin, enkephalins and ACTH, while stimulation at 200 Hz caused increases of serotonin (5HT) in the brainstem pain modulating pathways. Later, Dr. J.S.

Han showed that at 15-100 Hz dynorphin was released in the spinal cord, while enkephalins were released at 4 Hz.[11]. The mechanisms underlying the pain relief produced by TENS and EA appear to be complicated. Locally, at the site of application, electrical stimulation can increase vascular circulation, reduce edema and promote healing of various tissues. For example, bone repair can be augmented by applying an electrical current across the fracture, and muscular trigger points, a common source of pain seen in the clinical setting, can be observed to release after electrical stimulation. Centrally, it may release enkephalins, beta-endorphin, dynorphins, serotonin, noradrenaline for pain-relief and increase cortisol for an anti-inflammatory effect.

The TENS/ EA Endorphin Hypothesis

An integrated pain-modulating system as related to endorphins, hormones and neurotransmitters (Figure 1). Collectively, the TENS/ EA endorphin hypothesis may be divided into three levels of analgesia:

Fig. 1. Proposed mechanism of TENS and acupuncture (from Ph.D. thesis 1981 - Richard Cheng)

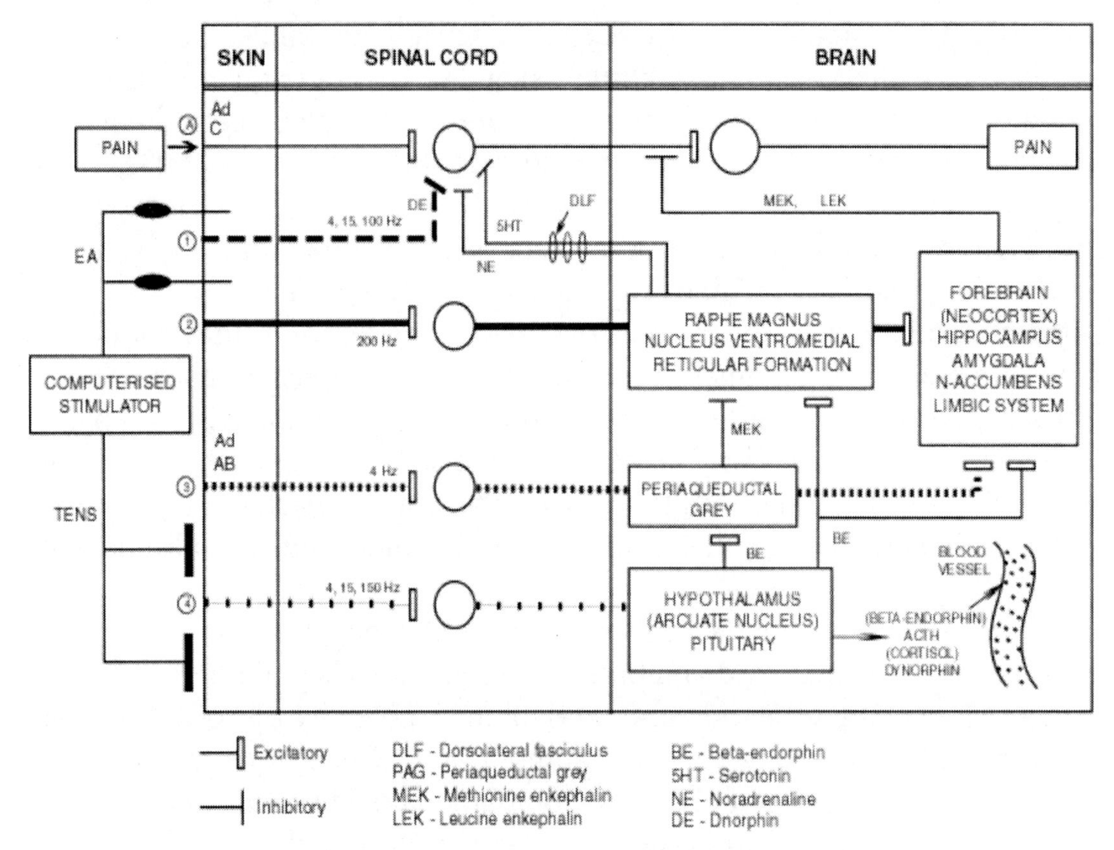

* The author use PCM - 8000, a programmable multi-frequency stimulator.

1. Local

TENS or EA, properly applied to an area of the body for a sufficient length of time, produces an area of numbness at the site. Such discretely localized analgesia is probably mediated by a segmental release of endorphins in the spinal cord which corresponds to the area of the body stimulated. Several lines of evidence have suggested that substance P is the neurotransmitter at the terminals of small primary afferent (A-delta and C) fibres responsible for peripheral pain transmission[13]. Immuno-histochemical analysis has further revealed that Methionine Enkephalin (MEK), one of the endogenous opioids, and substance P neurons have an intimate spatial relationship in the spinal cord (and also in the brainstem).[14] Moreover, it has been demonstrated that endorphins are able to suppress the release of substance P in the spinal cord.[15,16] Finally, TENS and EA can be shown to stimulate A-beta and A-delta peripheral fibres and that this releases endorphins in the spinal cord,[10] MEK being released at 4 Hz and dynorphin at 15-100 Hz (Figure 1- Pathway 1).[11,12]. Thus, a plausible mechanism exists to explain TENS and EA analgesia at the local/ segmental level.

2. Regional

Regional analgesia refers to pain-relief in a certain part of the body, e.g. upper arm, left upper body, etc. It may be due to the release of enkephalins in the midbrain nucleus (periaqueductal grey – PAG). The PAG has input to the brainstem nuclei (Raphe Magnus and Reticularis Magnocellulis) which can send down a descending inhibition through the dorsolateral fasciculus (DLF) to the spinal cord, where noxious input signals are then decreased[17]. It has been demonstrated that the Nucleus Raphe Magnus releases serotonin (5HT) while the Nucleus Reticularis Magnocellulis releases noradrenaline (NE) through the DLF.[18,19] The existence of regional mapping in these midbrain structures has been advocated, suggesting that stimulation of a certain midbrain area will cause analgesia in a corresponding area of the body.[17] At 4 Hz electrical stimulation can be shown to activate the PAG – brainstem nuclei pathway which can send down an inhibitory descending flow in the DLF and cause a regional analgesia.[12] There are also ascending projections for the midbrain nuclei which reach the hippocampus, amygdala, nucleus accumbens (limbic or emotional system) and neocortex (Figure 1 – Pathway 3). These also involve enkephalin release (MEK and Leucine Enkephalin or LEK) and are associated with memory, reward or sexual behaviour and higher pain associations. We can see that electrical stimulation of the PAG may modulate not only pain relief, but also impact some of its emotional, memory, reward and sexual behaviour. Evidence suggests that high frequency stimulation at 200 Hz may also stimulate the Nucleus Raphe Magnus and the nucleus reticularis magnocellularis directly, thereby activating the descending inhibitory pathway in the DLF, mediated by the release of 5HT and NA in the spinal cord[10,20] (Figure 1– Pathway 2).

3. General

Non-segmental analgesic effects can also be observed,[21] e.g. relief of facial, dental or headache pain by stimulation of distant areas on the hands, arms or legs. This with TENS and EA may be partially explained by beta-endorphin neurons in the Arcuate Nucleus of the hypothalamus and other endorphin neurons in the pituitary which can be stimulated by low frequencies (4Hz, 10Hz, 100Hz) TENS and EA[10] (Figure 1 – Pathway 4). Ascending neuronal pathways from the hypothalamus to the Nucleus Accumbens, PAG, Nucleus Magnus Raphe, Ventro-Medial Reticular Formation, Amygdala and Periventricular Nucleus of the thalamus could then activate widespread pain modulation.[22] Beta-endorphin and ACTH are made and released together into the CSF and blood stream by TENS or EA.[8] This may cause a generalized analgesia. The ACTH increases the synthesis of body cortisol which may add anti-inflammatory effects.[23] Hypothalamic and pituitary endorphins (beta-endorphin and dynorphin) are neural hormones which seem to provide particularly long duration analgesia of a generalized character.

ENDORPHIN RELEASE & OPIATE RECEPTORS

Frequency	Opiate	Opiate Receptors
4Hz	MEK LEK	Mu / Delta
15–100 Hz	Dynorphin	Kappa
4Hz	Beta-endorphin	MU
4 Hz	ACTH (↑ Cortisol Synthesis)	
200 Hz	SEROTONIN and NORADRENALINE	

Other scientific investigations and clinical observations review the functions of TENS or EA treatments.

Frequencies(Hz)	Functions	Possible Body Chemicals
1-10	Pain-relief Pleasure – learning	Met – Enkephalin Leu – Enkephalin
3	Euphoria, sexual	Dopamine
4	Pain – relief, anti-inflammatory	Beta – endorphin ACTH – cortisone
15	Pain – relief, addiction	B – dynorphin
100	Pain – relief, addiction	A – dynorphin
200	Pain – relief, insomnia, depression	Serotonin
300	Depression, pain - relief	Norepinephrine
1000 – 2000	Weight control Energy	Amphetamine
5 – 10	Tobacco – withdrawal	Nicotine
75 – 300	Sedative (minor tranquilizer)	Diazepam
75	Cannabis	LSD, THC
130	Auto – immune system	Immune system
3,3000	Headache	Autonomic nervous system
1000 – 5000	Interferential - decrease skin resistance - for deep organs	Internal organs
micro – current 300 – 600 micro A	Headache	Sympathetic nervous system
Acupuncture	Micro – current of damage: Pain – relief, healing, auto – immune system	Histamine, Bradykinin, prostaglandin, substance P long dendritic cells immune system

Fractal stimulations : restore the body to a healthy state

VIII. Fractal stimulations

Definition of fractal dimension: Temporal self-similarity - characteristics are unchanged under different scales of magnification and are called scale invariant.

Genetic algorithms are used to develop artificial intelligence for robots. If the cellular connections join in a fractal dimension, the robot will respond according to set parameters. In a fractal state, it allows for learning, memory and evolution; it allows life. Similarly, the human organs, like the ECG of the heart, or the EEG of the brain are all in a fractal state, indicating that they are healthy and alive. If the EEG or ECG deviates from the fractal state, they become random, chaotic or circular, rendering the subject sickly or diseased. As we grow older, we tend to deviate from the fractal state, becoming random or chaotic. We may have hypertension, arrhythmia, tachycardia, atrial fibrillation, dementia, epilepsy etc. If we can promote or push the EEG or ECG to a fractal state, we can become healthier. This can be accomplished by undergoing fractal TENS, fractal Qi-gong meditation or even fractal magnetic field stimulations. These treatments may prove to be useful in promoting better health and/or treat various illnesses.

The following diagrams describe the properties and ideas of fractal dimension:

In artificial intelligent experiment, imitated brain cells are programmed to connect randomly in the computer (Genetic Algorithm) Different intensities of stimuli may have inhibitions or facilitations on the outputs

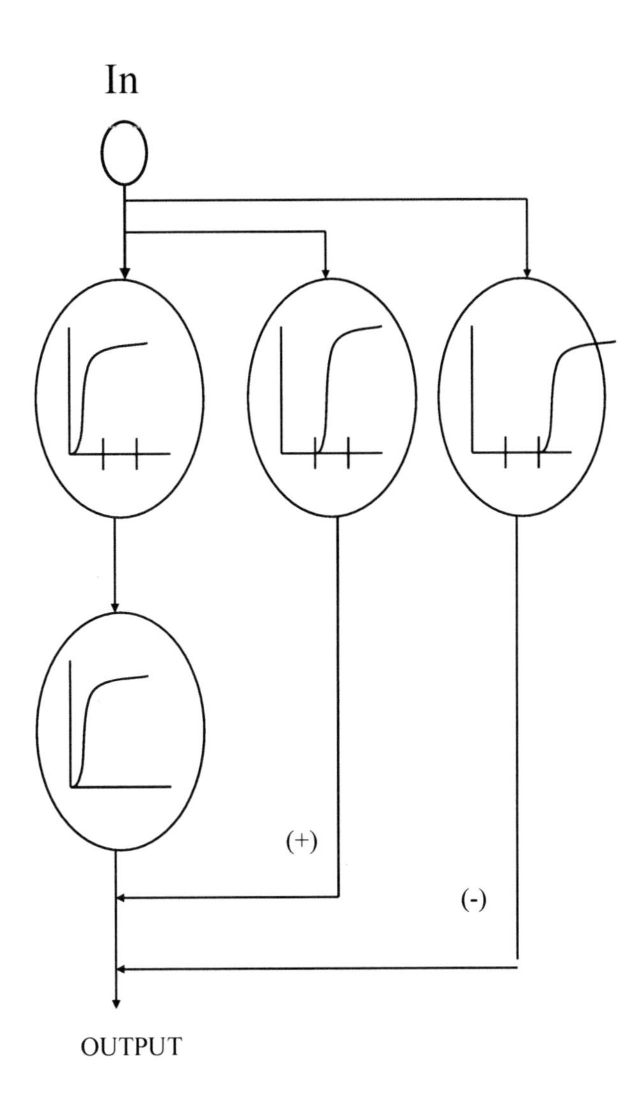

Genetic Algorithm: Cells connect randomly and only certain connections (fractal dimension) can response to the set parameter

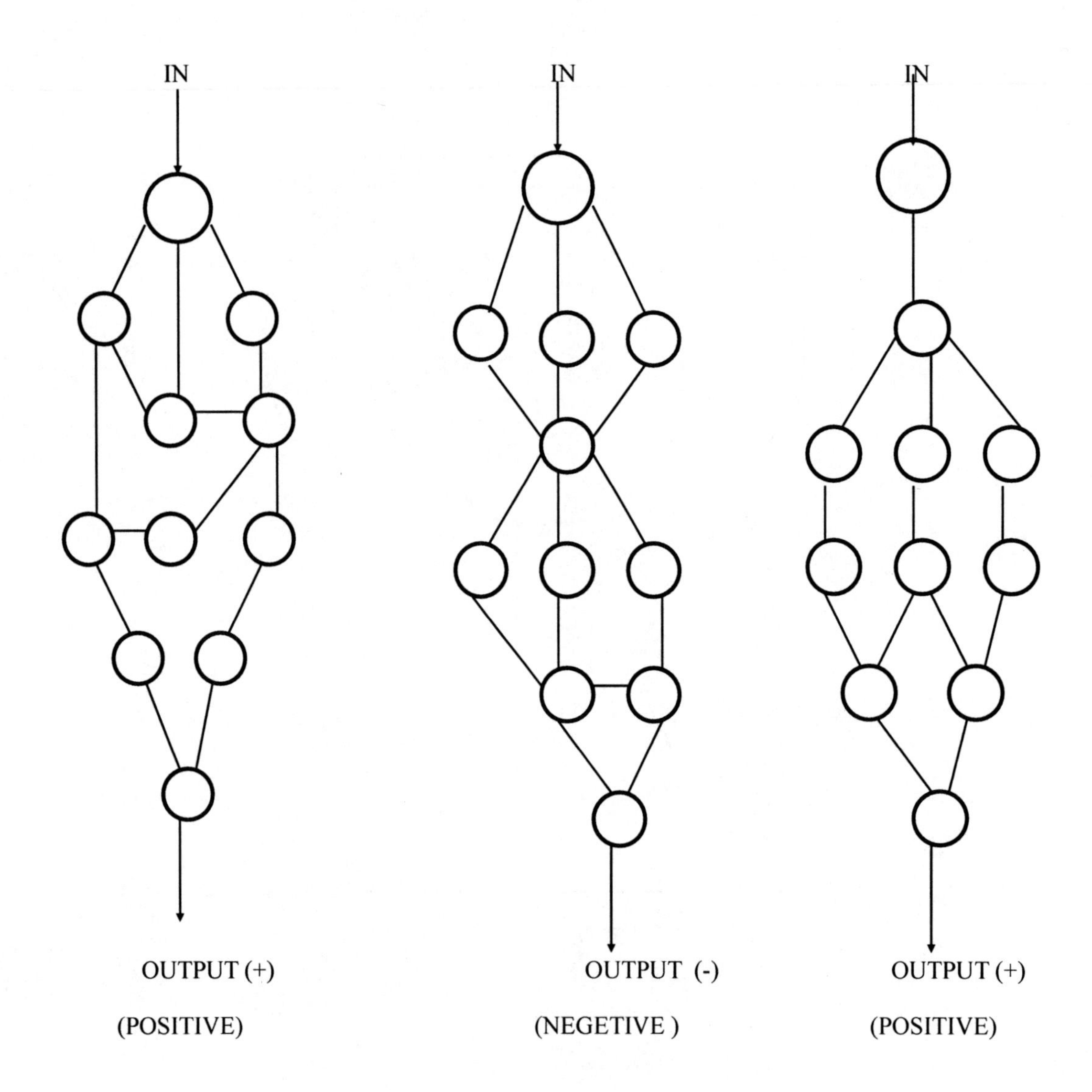

Two Dimensions :
Linear Dynamic and Non- Linear Dynamic

Three Dimensions :

Steady State
Periodic
Chaotic
Fractal

Fractal : definition

Temporal self-similarity :
characteristics are unchanged under different scales of magnification then the object is called scale invariant

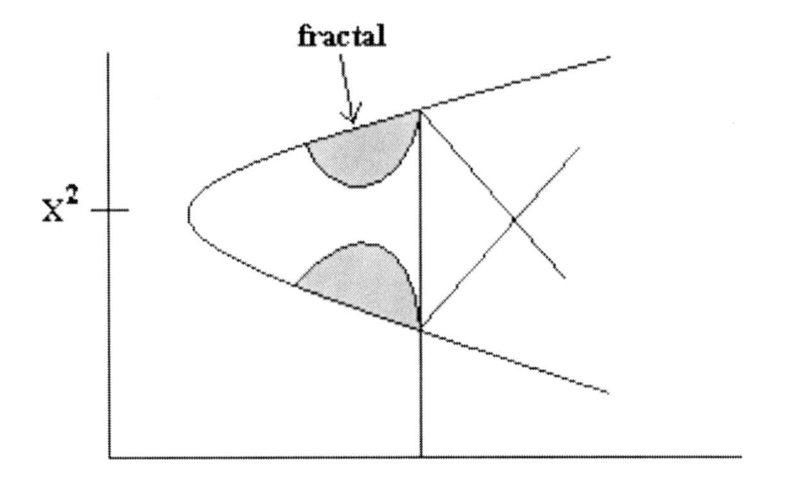

Fractal sits at the edge of chaos : $F(X^2) = X^2 + 1$

Fractals

Responding to environment adaptation

Can have Memory, Learning

Healthy Heart, Healthy brain

$F = L(r) = A\,r^{\alpha}$

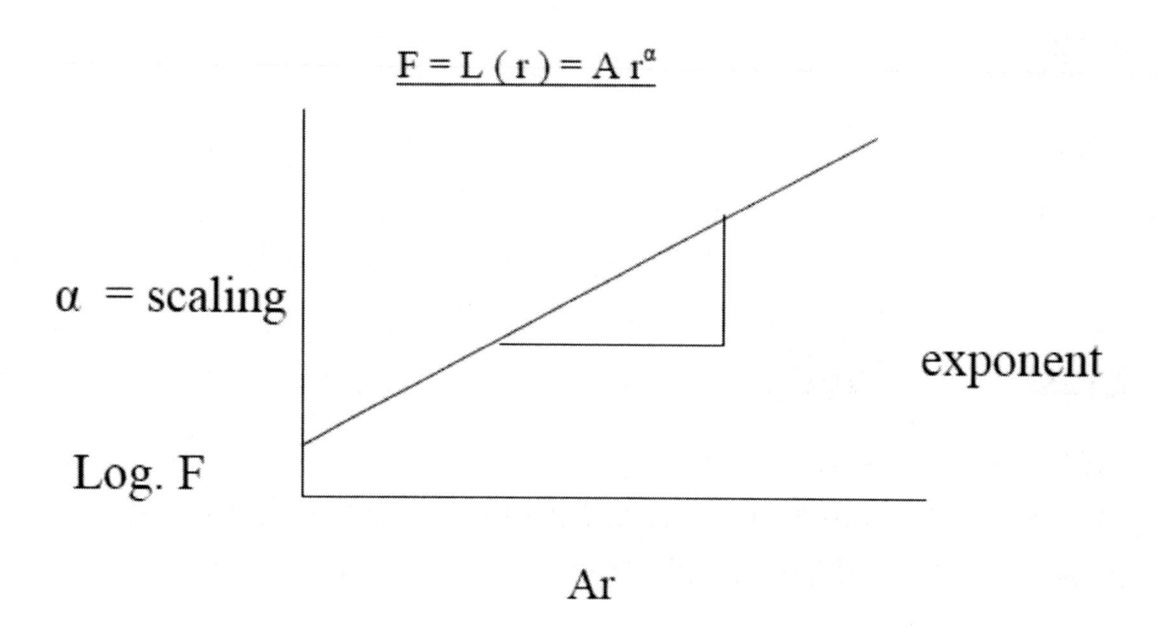

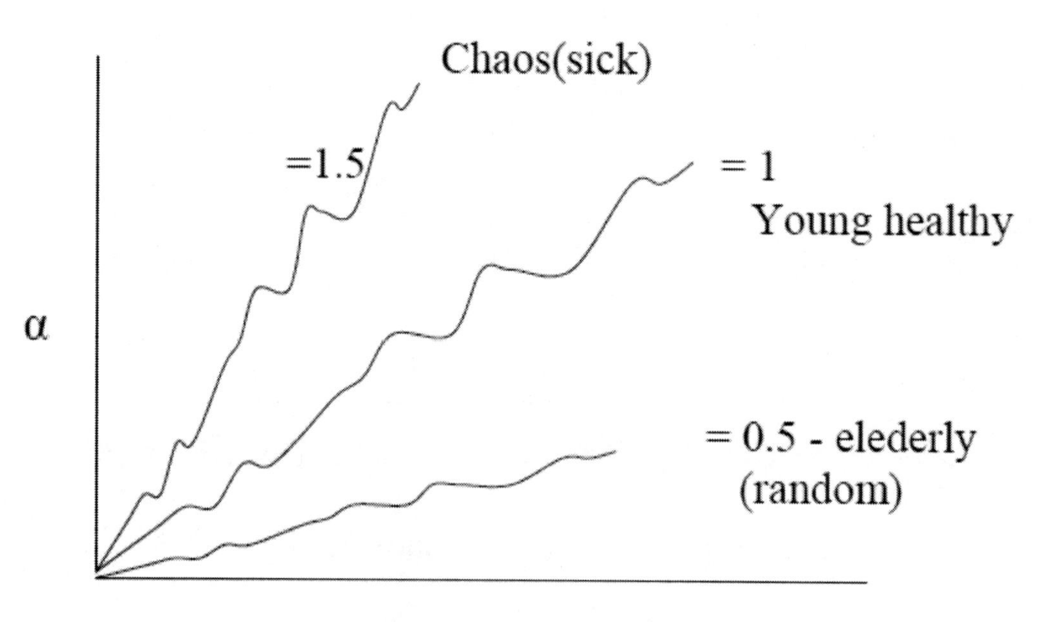

Fractal Gait Pattern

ECG

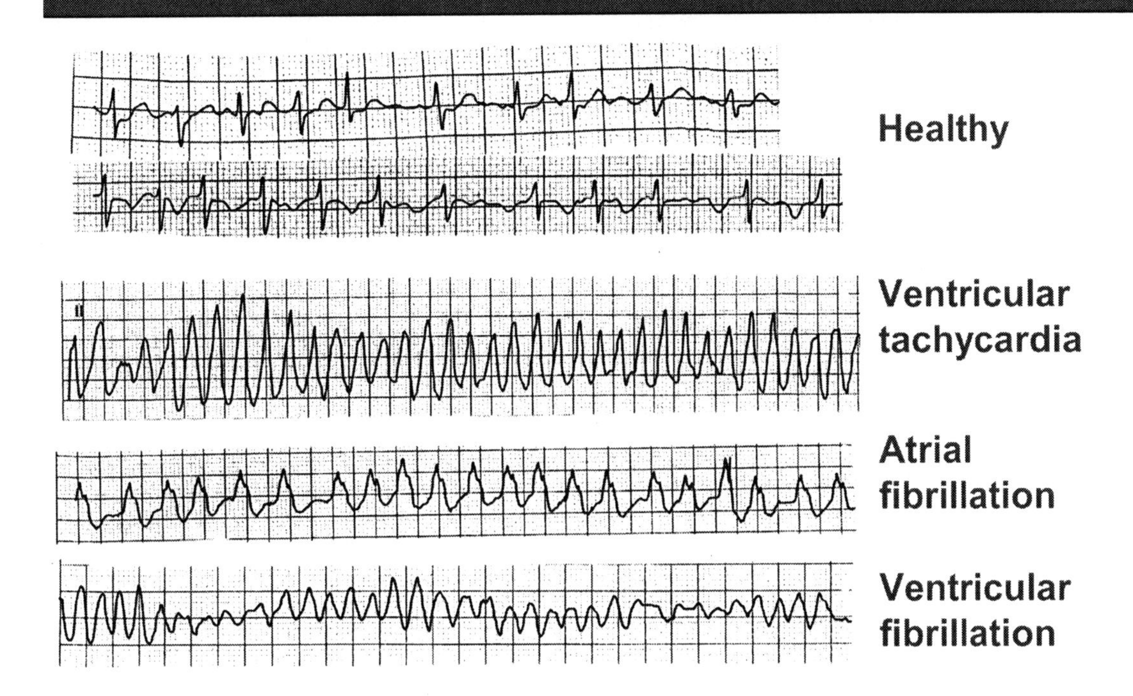
Healthy
Ventricular tachycardia
Atrial fibrillation
Ventricular fibrillation

EEG

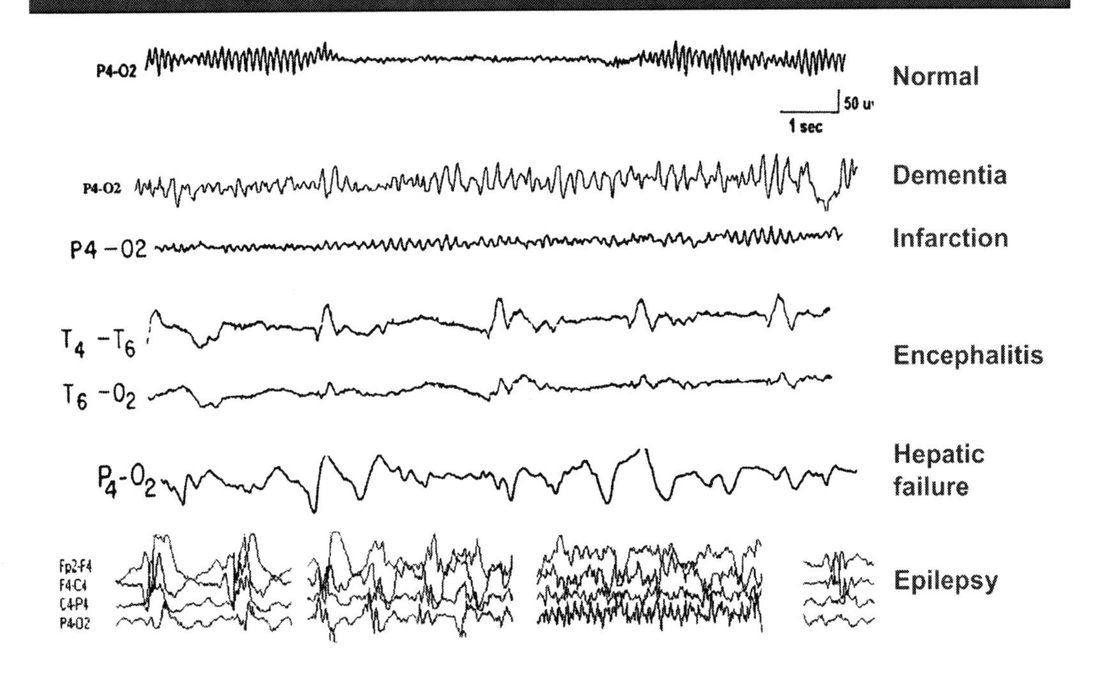
P4-O2
1 sec
Normal
P4-O2
Dementia
P4 - O2
Infarction
$T_4 - T_6$
$T_6 - O_2$
Encephalitis
$P_4 - O_2$
Hepatic failure
Fp2-F4
F4-C4
C4-P4
P4-O2
Epilepsy

The scaling component of temporal similarity ranges from 0.7 to 1.1 for healthy fractal state

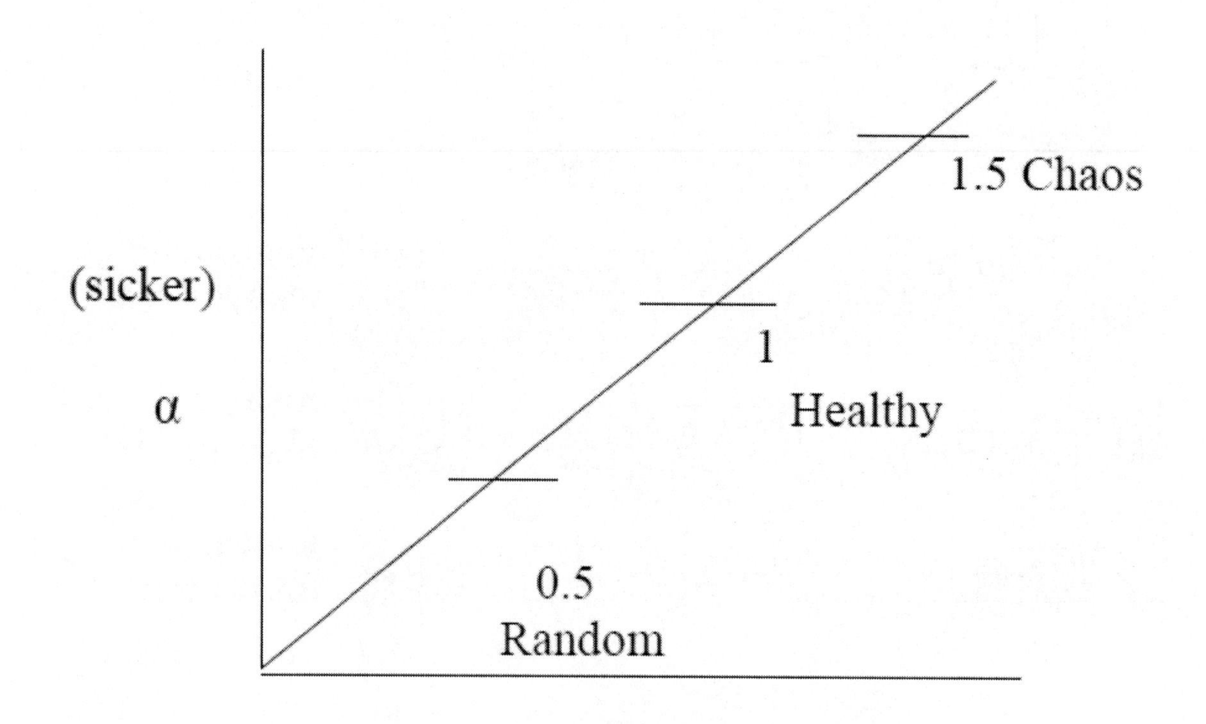

Healthy Dynamic Multiscale
(Long Range order)

<u>Healthy Heart, Brain and Others</u>

= (scaling component) 0.7 to 1.1

Promoting or stimulating

At the range 0.83

Discussion

Early childhood abuse or prolonged physical or emotional trauma will affect the networks and the memory of the brain cells and glia cells. They will activate stress hormones, cytokines (such as interferon, interleukins 1 and 6, tumor necrosis factor, etc.) neurotransmitters, as well as mal-functions of autonomic nervous system, immune system and over-active glia cells. It can entail some chronic diseases like chronic pain syndromes, depression, complex regional pain syndromes, fibromyalgia, heart disease as well as many auto-immune diseases such as multiple sclerosis, rheumatoid arthritis, lupus, scleroderma, Crohn's disease, psoriasis, hay fever, etc. Early treatment and prevention of tissue damage or stress may reduce the development of chronic pain syndrome and other illnesses. Regular exercise, Qi-Gong, manipulation, and TENS or acupuncture and fractal stimulations may be the best methods to reduce stress, prevent and treat these illnesses.

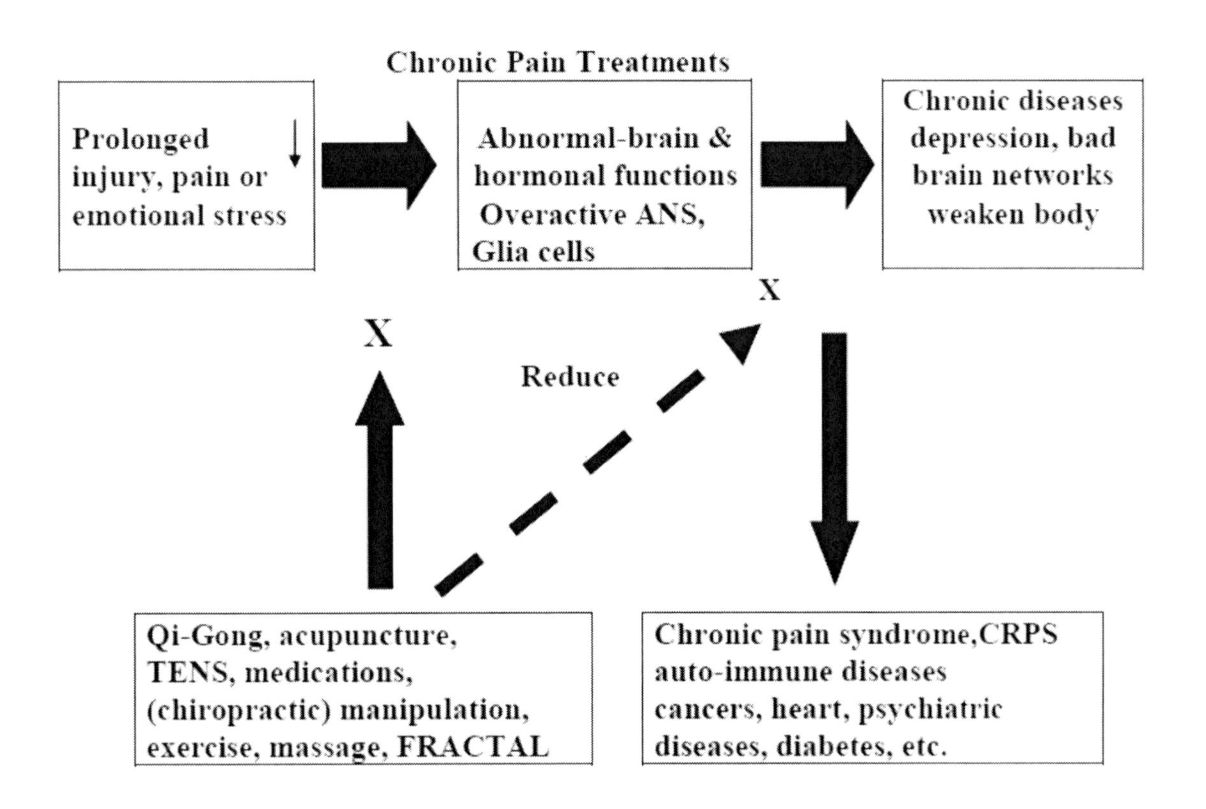

REFERENCES

1. "*Acupuncture Anesthesia*", People's publishing Co. Shanghai, China 1972.

2. Shealy, CN, Mortimer, J J and Reswich, JB: Electrical inhibition of pain by stimulation of the dorsal column: Preliminary clinical reports. Anesth Analg (Cleve) 45:489,1967.

3. Shealy, CN: Transutaneous Electroanalgesia: Surg Forum 23:419,1973.

4. Cheng, RSS and Pomeranz, B. Electrotherapy of chronic musculoskeletal pain: of electroacupuncture and acupuncture-like transcutaneous electrical nerve stimulation. Clin.J.Pain 2 (187) 143-149, 1987.

5. Kellaway, P: The William Osler Medal essay: The part played by electric fish in the early history of bioelectricity and electrotherapy. Bull Hist Med 20:112, 1946.

6. Stillings, D: A survey of the history of electrical stimulation for pain to 1900. Med Instrum 9:255, 1974.

7. Birch, J: *Essay on Electricity* ... by George Adams, ed 4. R Hindmarsh, London, 1772, p 519. Museum of Electricity in Life at Medtronic, Minneapolis, Minn.

8. Cheng, RSS (1977), *Electroacupuncture effect on cat spinal cord neurons and conscious mice: A new hypothesis is proposed, M.Sc. thesis*, Zoology, University of Toronto.

9. Mayer DJ, Price DD, Barber J et al. Acupuncture analgesia: Evidence for activation of a pain inhibitory system as a mechanism of action. In: Bonica JJ, Albe-Fessard D, eds. *Advances in Pain Research and Therapy.* New York: Raven Press 1976: 754-6.

10. Cheng RSS Pomeranz B. Electro-acupuncture analgesia could be mediated by at least two pain-relieving mechanisms: endorphin and non-endorphin system. *Life Sci* 1979;25:1957-62.

11. Han JS, Xie GX. Dynorphin: Important mediator for electro-acupuncture analgesia in the spinal cord of the rabbit. *Pain* 1984; 18: 367-76.

12. Cheng RSS. *Mechanism of Electro-acupuncture analgesia as related to endorphins and monoamines; an intricate system is proposed.* PhD Thesis. 1980, Zoology Department, University of Toronto.

13. Leeman, S, Substance P and Neurotensin, Eleventh International Congress of Biochemistry (Abstract), p. 541, Toronto. (1979).

14. Hokfelt, T, Ljungdahl, A., Terenius, L., Elde, R. and Nilsson, G., Immunohistochemical analysis of peptide pathways possibly related to pain and analgesia: enkephalin and substance P, Proc. Natl. Acad. Sci., U.S.A., 74, 3081-3085. (1977).

15. Mudge, AW, Leeman, SE and Frischbach, GD, Enkephalin inhibits release of substance P from sensory neurons in culture and decrease action potential duration, Proc. Ntl. Acd. Sci. USA, 76, 526-530. (1979).

16. Jessell, TM and Iversen, LL, Opiate analgesics inhibit substance P release from rat trigeminal nucleus, Nature (Lond.) 268, 549-551. (1977).

17. Basbaum, AI, Fields, HL, Endogenous pain control systems: brainstem spinal pathway and endorphin circuitry. *Ann Rev Neurosci* 1984; 7: 309-38.

18. Basbaum, AI, Clanton, CH and Fields, HL, Three bulbospinal pathway from the rostrol medulla of the cat: An autoradiographic study of pain modulating system, J. Comp. Neurol. 178, 209-224. (1978).

19. Yaksh, TL and Tyce, GM, Microinjection of morphine into the periaqueductal gray evokes the release of serotonin from spinal cord, Brain Res., 171(1), 176-181. (1979).

20. Cheng RSS, Pomeranz, B. Monoaminergic mechanisms of electroacupuncture analgesia, *Brain Res.* 215, No. 1-2, 77-93. (1980).

21. Chapman, CR, Chan, AC and Bonica, JJ, Effect of inter-segmental electrical acupuncture on dental pain: Evaluation by threshold estimation and sensory decision theory, Pain 3, 213-227. (1977).

22. Barchas, JD, Akil, H, Elliot, GR, Holman, RB and Watson, SJ, Behavioral neurochemistry: neuroregulators and behavioral state, Science, 200, 964-973. (1978).

23. Cheng, R, McKibbin, L, Buddha, R, and Pomeranz, B, Electroacupuncture elevates blood cortisol levels in naïve horses; sham treatment has no effect, Intern. J. Neurosci. 10, 95-97. (1980).

24. Melzack R. From the gate to the neuromatrix; Pain, suppl. 6, 121-126. (1999).

25. Llinas R, Ribary U, Joliot M and Wang XJ, Content and context in temporal thalamocortical binding. In *Temporal Coding in the Brain* (eds Buzsaki G, Llinas R., Singer W., Berthoz A. and Christen Y.) p251-272, Springer, Berlin. (1994).

IX. PAIN SYNDROMES AND TREATMENT METHODS

Peripheral Nerves

1. A delta and C – small non-myelated fibres which are noiceptors for the skin, muscle, tendon, fascia, joints, cornea and tooth pulp. A-beta fibres are responsible for touch and A-alpha fibres are for vibrations and proprioceptions.
2. Visceral pain receptors – mostly responsible by sympathetic afferent fibres. A delta and C fibres may partly responsible for cardiac pain.
3. Inflammatory pain may be mediated or potentiated by 5 – hydroxytrytamine, histamine, bradykinin, and prostaglandins.

A) Excitatory Neurotransmitters

Acetacholine, glutamate, tachykinnins – substance P and neurokinin A, dynorphin is antianalgesic – in certain concentration ranges.

B) Inhibitory Neurotransmitters

i) **Glycine**

ii) **Gamma** (GABA): aminobutyric acid
- GABA A - benzodiazepine
- GABA B - baclofen

iii) **Endorphins**
- Beta–endorphin
- Dynorphin
- Methionine enkephalin
- Leucine enkephalin

Prolonged activation of opiate receptor by exogenous opioids causes increase in protein kinase C, which interferes with action of opiate receptor and increases effect of N-methyl-D-aspartate (NMDA) receptor activation, resulting in tolerance and hyperalgesia.

iv) **Serotonin**: Analgesic and anti-depressant effect, involves in descending pain control from the midbrain and brain stem. Serotonin projecting neurons from raphe nucleus to the prefrontal cortex is involved in the control of depression.

v) **Nor-epinephrine**: involved in descending inhibitory pain control.

vi) **Adenosine receptors** may play a role in spinal opioid analgesia induced by transcutaneous electrical nerve stimulation (TENS) or electro-acupuncture.

A) SPECIFIC PAIN SYNDROMES

1. **Neck and back pain** - radiculopathy

2. **Joint pain** - arthritis and degeneration

3. **Myofascial pain syndrome** - fibromyalgia, tendonitis, fascitis, trauma.

4. **Peripheral neuropathies and neuralgia** – trauma, infections (e.g., acute herpes zoster)

5. **Facial pain** - trigeminal neuralgia, Bell's palsy

6. **Headaches**

7. **Chronic pelvic pain**

8. **Central pain syndromes**

9. **Painful medical diseases**

10. **Cancer pain**

11. **Complex Regional Pain Syndromes I and II**

Complex Regional Pain Syndrome (CRPS)

(Reflex Sympathetic Dystrophy – RSD) The presence of regional pain and sensory changes following an injury, together with associated regional findings such as abnormal skin colour, temperature change, sweating and tissue swelling, spontaneous pain – allodynia, muscle wasting, joint stiffness and loss of functions.

Two Types of CRPS
Type I - RSD without a definable nerve injury
Type II - Causalgia with a definable nerve lesion

Signs and Symptoms

- Spontaneous pain – allodynia, hyperalgesia
- Abnormal skin color, temperature or sweating – cold, hot or cyanotic, pale
- Swelling (edema)
- Muscle wasting, joint stiffness, loss of function
- Osteoporosis
- Headache, dizziness, tinnitus
- Skin – cold, deep structures – hot
- Abdominal pain, nausea, vomiting , diarrhea, cramps, weight loss
- More are female
- Sleep disturbances
- Depression
- Fatigue and fibromyalgia

Involvement

1. Cervical spine – neck pain, headache, nose bleeding, dizziness, tinnitus, TMJ pain.
2. RSD of the blood vessels to kidney - Increased blood pressure – Back Pain - Temporary bleeding – Headache and dizziness, etc.
3. Celiac area – abdominal pain peptic ulcer and vomiting, weight loss.
4. Superior and inferior mensenteric plexi – diarrhea, abdominal cramps and weight loss.
5. Cardiac plexus – chest pain, abnormal heart beat, tachycardia, and heart attack.
6. Carotid and vertebral plexi – severe vascular headache, dizziness, tinnitus, syncope.

Theories for Pathogenesis
Injury results in an inflammation and irritation of the sympathetic nerves from the end organ to the spinal cord and can spread to the other autonomic nervous system. Inflammation and irritation causes constriction of the blood vessels and release of substance P under the skin. The overactive sympathetic nervous system may be due to the overactive glia cells in the brain. These cause hyperalgesia, allodynia, red reflex, hot and cold, ischemia, sweating, muscle spasm and wasting, hair loss and osteoporosis.

Investigations
Diagnosis is often made on history, signs and symptoms, and elimination of differential diagnoses by various tests.
- Nerve conduction test – nerve injury.
- Thermogram – ischemia
- Bone scan – osteoporosis
- X-ray, US, CT scan, MRI to eliminate other possible causes.

Treatments
- TENS – home unit for daily use
- Acupuncture
- Trigger point injections
- Nerve blocks
- Meditation
- Physiotherapy – low impact exercise
- Fractal stimulations

Medications
- NSAIDs, Toradol, Voltarens, etc.
- Topical NSAIDs & cortisol or EMLA (anaesthetic) creams
- Antidepressants, e.g., nortriptyline
- SSRI, e.g., floxetine
- Anticonvulsants, e.g., gabapentin, pregablin
- Minor tranquilizers, e.g., Diazepam
- Narcotics, e.g., Codeine, Oxycodone,
- Morphine, Durgesic patch
- Cannabis, e.g., Marinol, Nabolone)

Surgery
SPA (Stimulation Produce Analgesia) by implanting electrodes in the spinal cord or brain.

Prognosis
Early treatments with TENS, acupuncture or nerve block may result in better pain relief, improvement or even remission. Chronic pain may be helped and or go into remission with long-term TENS, meditation and a strengthening exercise program. Some patients may need long-term psychiatric care and analgesics.

B) TREATMENT METHODS

Medications

1. Topical – Pain control creams
   Capsaicin + menthol + methyl salicylate
   Capsaicin + nitroglycerin
   NSAIDs – methyl salicylate, voltaren.
   Anaesthetic – EMLA xylocaine 5-10%
   Herbal ingredients – camphor, menthol, tumorist, radix notoginseng.

2. Acetaminophen – do not exceed 4 gms per day for normal people; can have fatal liver toxicity especially for hepatic and renal cardiac impairment.

   NSAIDs – ASA – Mobicox, Celebrex.
   Cyclo oxygenate (Cox – 1 and – 2)
   - Celecoxib, Nofecoxib, Meloxican
   - Naproxen, Relafen, Arthrotec, Ibroprofen, Diclofenac,
   - Toradol, etc.

Neuropathic Pain

- Co-analgesic
- Moderate to severe pain in combination & opioid

Antidepressants

- Tricyclic antidepressants: Serotonin and Nor-epinephrine uptake block sodium channels, adenosine receptors & N–methyl–D–aspartate (NMDA) receptors.

- Amitryptaline (10-75 mg qhs)
- Desipramine

- Nortriptyline (10-75 mg)
- Venlafaxine
- Paroxetime

Anticonvulsants

- Gabapentin (100-2400 mg/day)
- Lyrica (75 mg bid to tid)

Cannabis

- Nabolone 1mg to 2mg per day

Co–analgesics

Muscle relaxants

- Baclofen – gamma – amino – butyric acid (Gaba)
- Cyclobenzapine
- Carisopsodo; and methocarbamal
- Orphenadrine
- Minor sleeping tranquilizer, e.g., Valium, Lorazepam, Ativan, etc.
-

Trigger Point Injections

- xylocaine + marcaine +/- steroids

Nerve Blocks or Epidural Nerve Blocks & Steroids

Opioids (Narcotics)

Opioid Receptors	
Mu	Analgesia, Euphoria
Delta	Analgesia, Euphoria
Kappa	Psychotomimetic, analgesia
Sigma	Affect change, psychotomimetic
Morphine 6 – glucuronide (m-6-G)	Analgesia
Methadone: NMDA receptor blocker	No narcotic induced hyperalgesia

Narcotics

Opioid	Dose	Duration
Morphine	10 mg	3 – 4 hrs
Hydromorphone	3 mg	3 – 4hrs
Meperidine	30mg	3 – 6 hrs
Methadone	10 mg	4 – 8hrs
Oxycodone	10 mg	3 – 6 hrs
Transdermal	25 µg / hrs	72 hrs
Codeine	30 mg	3 – 4 hrs

Opioids

Oral CR/SR	Starting Dose	Indications	Titration Time
Codeine contin	50 – 30 mg q 12hrs.	Mild to moderate pain	48 hrs
Morphine MS contin	12 – 30 mg q 12 hrs	Severe pain	48 hrs
Oxycodone oxycontin	10 mg q 12 hrs	Severe pain	24 hrs
Hydromorphone (Hydro morph contin)	3 mg q 12 hrs	Severe pain	48 hrs
Fentanyl Durgesic patch	25 µg 1 hrs	Severe pain	48 hrs
Methadone	10 mg	Chronic pain	

	Oral	Parental
Morphine sulphate	30 mg	10 mg
Codeine	200 mg	120 mg
Meperidine	300 mg	75 mg
Oxycodone	30 mg	15 mg
Propoxyphene	100 mg	50 mg
Hydromorphine	7.5 mg	1.5 mg
Antagonist		
Nalbuphine		10 mg
Butorphanal		2 mg

The WHO Analgesic Ladder

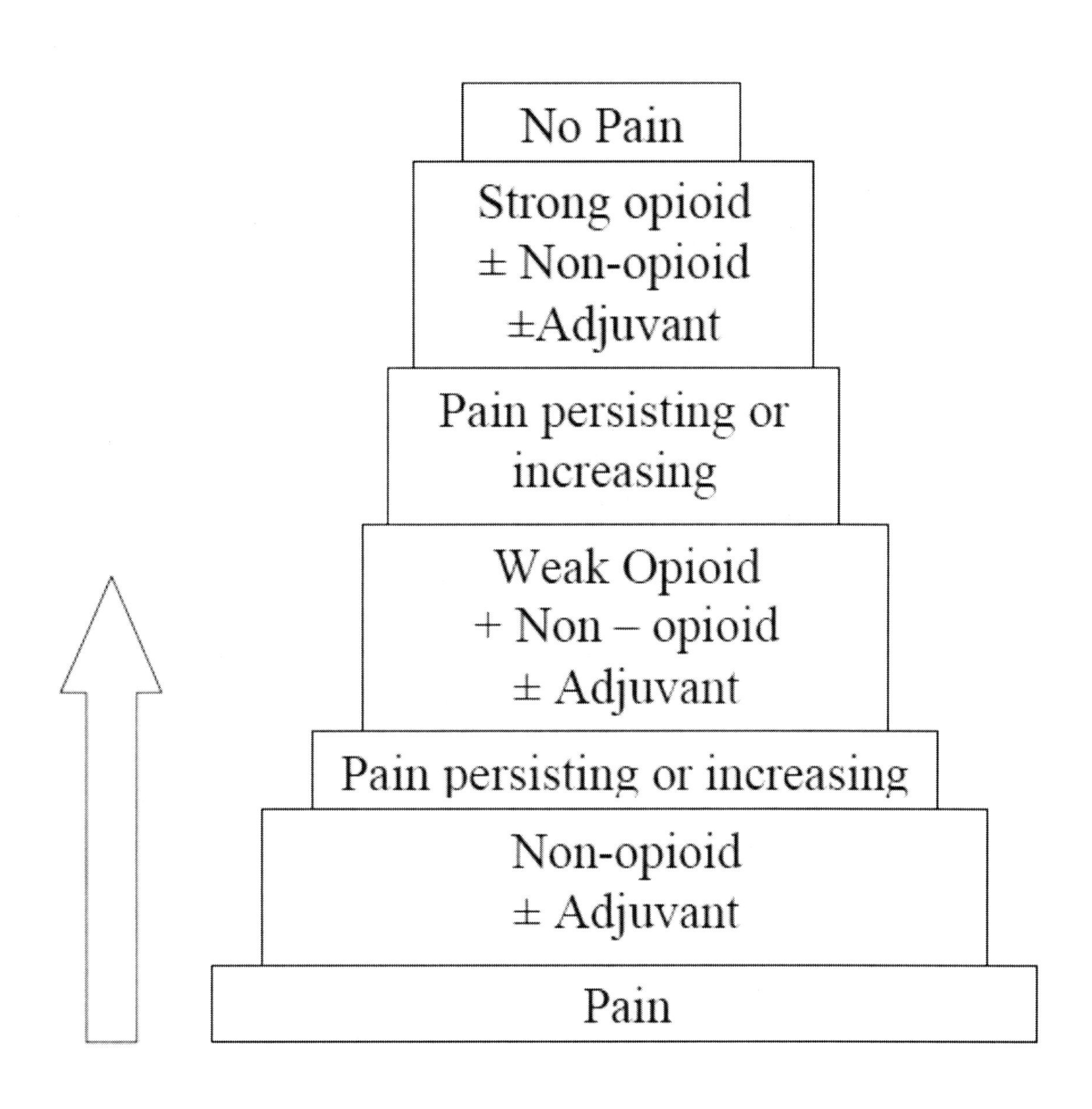

Non – Pharmacological Treatments of Pain

A. Good for mild to moderate pain and long term chronic pain treatment.
B. May have synergistic healing or strengthening effect for certain illnesses.
C. Fewer side effects, especially avoiding long-term drug toxicity or addictions.

1. **TENS (Transcutaneous Electrical Nerve Stimulation) and EA (Electro-Acupuncture)**

2. **Heat or Cold Compresses**
   Heat therapy is generally good for joint stiffness, spasm, abdominal or menstrual cramps. Cooling therapy is good for reducing edema, hyperemia, inflammation and pain, such as frozen shoulders, rheumatoid arthritis, herniated disc, trigger zone pain, and epicondylitis.

3. **Exercise Therapy** is usually beneficial for many chronic pain syndromes, especially those of the musculoskeletal system. Stretching and strengthening exercises (with or without weights or equipment) are more useful. Any exercises, especially vigorous ones that cause more pain and injury, should be avoided by patients.

   Massage Therapy and Manipulation
   Soothing feelings and pain reduction may be achieved especially for muscle pain, tendonitis and other connective tissue problems.

   Traction and Orthoses
   Traction is often very effective and useful for neck and back pain due to herniated disc, degeneration, spondylolisthesis, etc.
   Neck and lumbar supports by orthoses may reduce pain.

   Meditation and Qi-Gong
   Mindful meditation is found to be useful in treating chronic pain and many diseases.

4. Fractal stimulations

X. CONTRA-INDICATIONS AND PRECAUTIONS FOR ACUPUNCTURE AND TENS

1. To prevent the transmission of disease (especially hepatitis and meningitis), use only sterilized and disposable needles.
2. Avoid unnecessary use of needles or TENS on pregnant women or young children.
3. Know your anatomy before inserting the needles. Puncturing the lung can cause pneumothorax; puncturing the spleen, liver or kidney can cause a small hemorrhage. Avoid needling of tumour sites, skin infections and hemorrhagic diathesis such as hemophilia.
4. Do not apply current across the heart area if the patient has heart disease or a pacemaker and do not stimulate the carotid sinus to cause syncopy.
5. Fainting - During acupuncture treatment, some patients may develop symptoms of dizziness, vertigo, palpitation, nausea, cold sweating, pallor or even loss of consciousness. In this case, the needles should be removed immediately and the patient allowed to lie flat. If the symptoms persist, press or puncture point CV26 or P 9 with a fingernail or needle. Often the patient will recover quickly, otherwise emergency measures will have to be taken. N.B. To prevent the occurrence of such symptoms, it is advisable to have patients lie down while giving them treatment.
6. Stuck needle – Spasm of the muscle may hold the needle tightly. Massaging the surrounding area to relax the muscles or allowing the patient to calm down for a few minutes will allow the needle to be easily removed.
7. Bent and broken needles – Avoid using bent or cracked needles. Bent needles can be removed gently by following the direction of the bend. Broken needles are removed either by hand, forceps or even surgery
8. In rare instances a patient may find that electrical stimulation is very unpleasant and intolerable in which case s/he should be excluded from further use of the stimulator.
9. Do not use TENS whenever pain syndromes are undiagnosed, especially before consulting a physician or until etiology is established.
10. Do not place electrodes in a manner that will cause current to flow transcerebrally.
11. Do not connect lead wires to electric power cords or AC outlets.
12. TENS devices should be kept out of the reach of children.

XI. ANALYSIS OF ACUPUNCTURE POINTS

It is admirable that the Chinese discovered and designed the meridians and acupuncture points over the surface of the human body more than two thousand years ago, considering that they had very little knowledge of anatomy and physiology at that time. Perhaps the discovery was based purely on clinical and factual observation. The mapping out of acupuncture points was possible only after an enormous amount of clinical experience. Careful observation reveals that the names of the fourteen classical meridians do not necessarily correspond to the physiological function of the organs to which they refer. However, there has been virtually no revision of these classical meridians for over two thousand years. Instead, a great number of acupuncture points have been added, which are considered extra-meridian points. Because of some of the dramatic clinical results, acupuncture points have always been looked upon as points of mystery. So, exactly what are acupuncture points?

In general, an acupuncture point can be defined as an effective site of physical stimulation in the human body. This effective site can be in one of many different tissues, namely nerve, muscle, vascular structure, periosteum and even mucosa.

1. Nerves

a. **Cranial nerves** – can be easily reached on the head, face and the external ear. On the face, the trigeminal and facial nerves are the most common cranial nerves to be stimulated. In the external ear, there are at least three or four branches of the major cranial nerves. The location of an acupuncture point on a cranial nerve is always either at the nerve trunk where it leaves the foramen or at the very end of its terminal branch. Acupuncture points located on cranial nerves are always at the site of anastomosis of two different nerves or at the point at which a nerve branches bilaterally.

b. **Peripheral nerves** – Acupuncture points found on peripheral nerve trunks in the upper and lower extremities are always immediately above or below the bone head. These points are called supracondylar or infracondylar points. Points on peripheral nerves may also be interosseous, e.g. between the ulna and radius or the fibula and tibia. Peripheral nerve points are also located at the site of anastomosis of the terminal branches.

c. **Spinal nerve roots** – In the posterior aspect of the body, along the spinal column, from the neck down to the tail, acupuncture points are located on the posterior primary ramus of almost every spinal segment. Again, the points are at the anastomosis of the spinal nerve roots at the midline of the spine.

2 . Muscle

There have been suggestions that acupuncture points are motor points. A motor point is defined as a point in the muscle which, when electrical stimulation is applied, will produce a maximum contraction with minimal intensity of stimulation. The motor point lies close to the point at which the nerve enters the muscle, and it approximates but is not identical to the end plate zone of motor nerve endings. The exact location of the motor may vary from person to person, but is usually a fixed anatomical site. Acupuncture points are designed at the musculotendinous junction. An obvious example is the point just proximal to the tendon insertion on a bony prominence, and therefore is known as supratendinous insertion points. Acupuncture points sometimes are located between the bellies of two muscles and are called intermuscular points.

3. Vascular Structures

The dramatic clinical results obtained by acupuncture therapy in the sympathetic nervous function are still not fully understood. Examination reveals that the terminal arterial structures in the extremities are the site of some points most influential on the sympathetic function. Typical examples are L14 in the upper extremity and Liv3 in the lower extremity. Arteries in the extremities travel very close to the major peripheral nerves. When they reach the terminus of the hand or foot, however, they form double arches. There appears to be one point on the arterial arch which, when stimulated, is very effective in altering the status of the sympathetic system of the whole body, as all blood vessels are richly innervated by the autonomic nerves.

Acupuncture points which have obvious influence on the systemic blood pressure are on the arteries at the interosseous and infracondylar points – St 36 and LI10. There is one particular acupuncture point designed on the carotid sinus in the neck. For hypovolemic shock, acupuncture therapy using this point is very effective. These effective sites are mostly on the capillary interface in the fingertips, toes and in the midline just above the lips. Some acupuncture points are right on certain sympathetic ganglia. Scalp acupuncture and osteoacupuncture are done by stimulating the periosteum in the skull, hands and feet.

4. Periosteum
There are a lot of sensory receptors at the periosteum. Stimulation of the periosteum can be very painful and can create a strong sensory input to the brain. Probably Aδ and C fibres are stimulated by this method.

5. Skin
Can acupuncture be skin deep? In Europe, particularly in Austria, some acupuncturists are inserting needles very superficially into the skin surface of the human body. The distribution of nerves on the skin surface is called a dermatome. These areas are richly supplied by free nerve endings or sensory receptors which are responsible for the sensory inputs, such as touch and pain. Some of the acupuncture points are specifically right at the cutaneous area overlying the fasciae. This probably stimulates mostly the Aδ and Aβ fibres.

6. Mucosa
Very few acupuncture points are located on the mucosa.

Conclusion

The acupuncture EA and TENS stimulate mostly the Aβ and Aδ fibres and less frequently the Aα and C fibres. This can send a message to the spinal cord and brain to release endorphins and other hormones. The insertion of the needles to the acupuncture points creates cellular damage and micro-tissue injury. Thus cellular chemicals such as histamine, prostaglandins, etc., are released locally. These create a micro-current for which the skin resistance will drop at the site of tissue damage. The tissue repair reaction will begin and can last from 3-10 days. This can affect the fibrinolytic system, coagulation system, immune-complement system and white blood cell counts. Thus the acupuncture may have an after-effect for 3-10 days. The main idea is to stimulate the nerves to induce the release of endorphins for pain relief, regardless of whether or not it is an acupuncture point.

XII. TECHNIQUES OF ACUPUNCTURE, EA AND TENS

Patients should be in a comfortable position for acupuncture, EA or TENS treatment (Figure 10). They should be allowed to lie in a supine position, especially if they have syncope with needle insertions. Mechanical or electrical stimulations are shown in Figure 11.

Techniques of Acupuncture, EA and TENS
Figure 10.

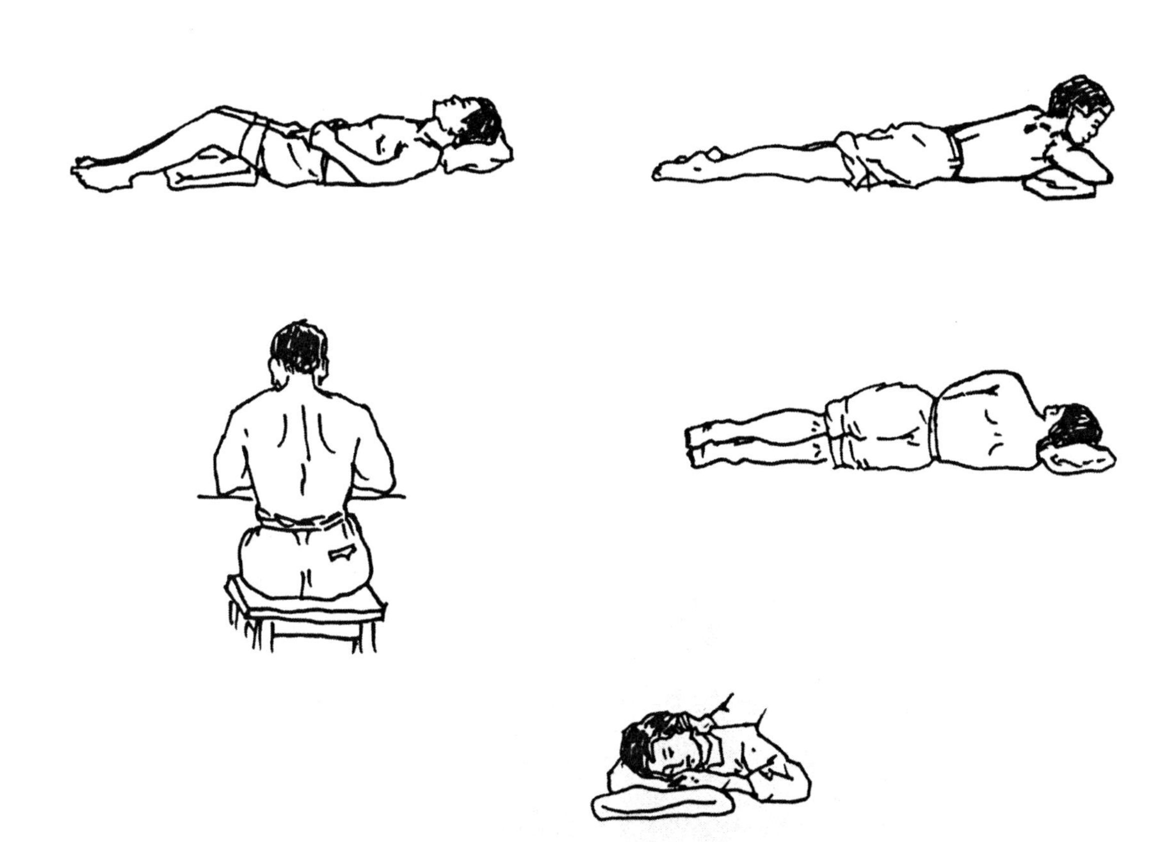

Techniques of Acupuncture, EA and TENS
Figure 11.

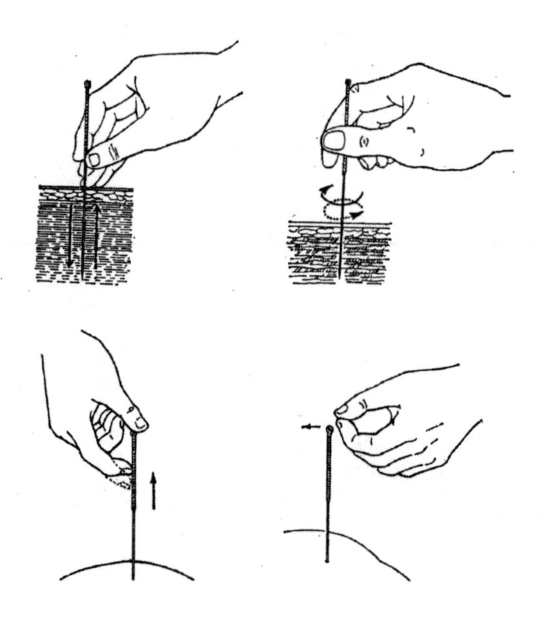

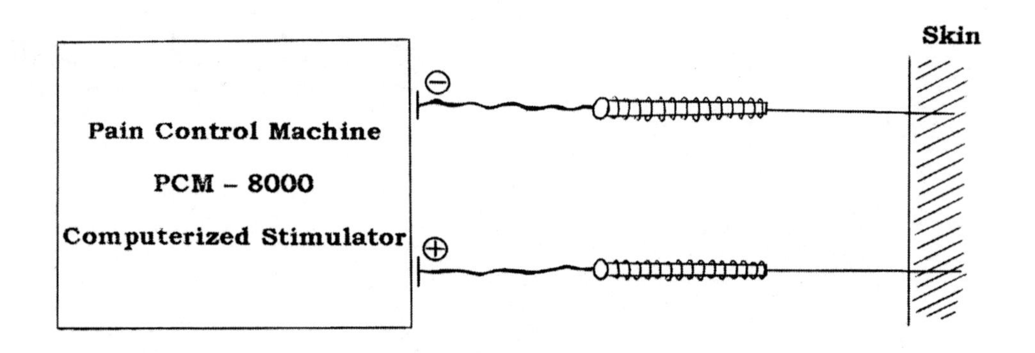

TENS

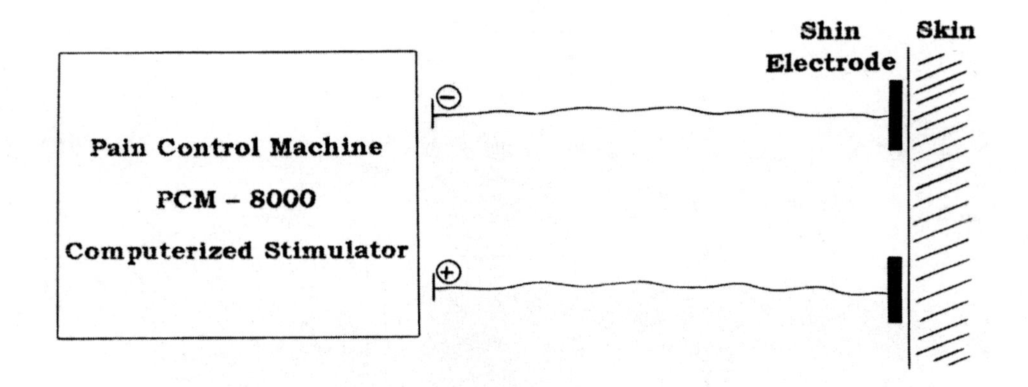

XIII. CLINICAL TREATMENTS

1. Headache

Dr. John Edmeads describes the Seven Danger Signals for Headaches suggesting the possibility of serious disease that may require further investigations including CT scan and MRI.

1. Failure of the headache to conform readily to an innocuous pattern.
2. Onset of headache in childhood or in or after middle age.
3. Recent onset and progressive course.
4. Other neurological or general symptoms.
5. The patient "looks sick" or "doesn't feel right".
6. Abnormal physical signs
7. Meningeal irritation…….the cost of missing it is so great.

Classification of Headaches

A. Vascular headache

(i) Migraine, (ii) Cluster headache

B. Muscular headache

(i) Tension headache, (ii) Muscular headache

C. Neuralgic headache

(i) Occipital neuralgia, (ii)Trigeminal neuralgia, (iii) Facial neuralgia

D. Psychogenic headache

E. Organic headache

F. Headache due to altered intracranial pressure

Acupuncture can be very effective in treating A, B, C and D headaches.

Acute headache – local trigger acupuncture points plus associated distal points.

Prophylactic treatments for 10 times – distal points (2-3 times per week).

Treatment of Headache
Table 1
Psychogenic Headache Points

Acupuncture Points	Location	Nerve
He 7	Lateral to the flexor carpi ulnaris – in the wrist crease	Ulnar nerve
Gv 14	Mid-point between C. 7 spinous process and T. 1 spinous process	T.1 Nerve root

Treatment of Headache
Diagram 1.
Psychogenic Headache

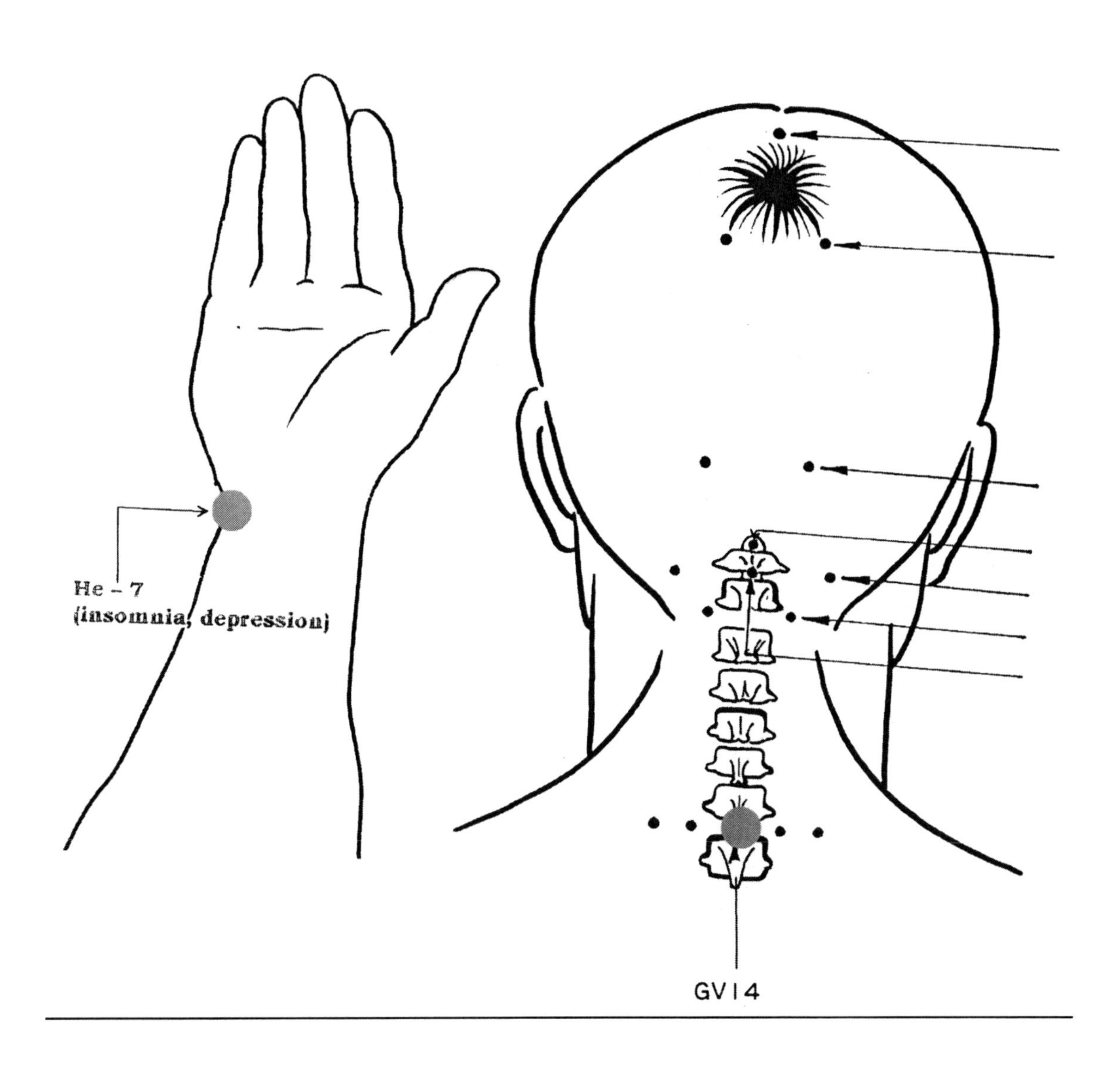

Treatment of Headache
Table 2. Sympathetic Switches in the Extremities

Acupuncture Point (Fig)	Location	Artery
LI 4	Between 1st and 2nd metacarpals midpoint of the 2nd metacarpal	Dorsalis radialis artery - entering the dual arterial arch
LI 11	Midpoint between biceps tendon and lateral epicondyle - on the elbow crease	Radial recurrent artery
Liv 3	Between 1st and 2nd metatarsals - proximal to the heads	Dorsalis pedis artery
St 36	3" below the inferior margin of the patella, 1" lateral to the anterior tibial ridge	Deep peroneal artery

Treatment of Headache
Table 3. Panic Switches

Acupuncture Point (Fig)	Location	Arterial Structure
GV 26	Upper one third in the labial groove, upper lip	Arterio-venous
He 9	Medial lower corner of nail bed, little finger	Anastomosis Arterio-venous
P 9	Lateral lower corner of nail bed, middle finger	Anastomosis Arterio-venous
Lu 11	Medial lower corner of nail bed Thumb	Anastomosis Arterio-venous Anastomosis

Treatment of Headache
Diagram 2.
Distal Points

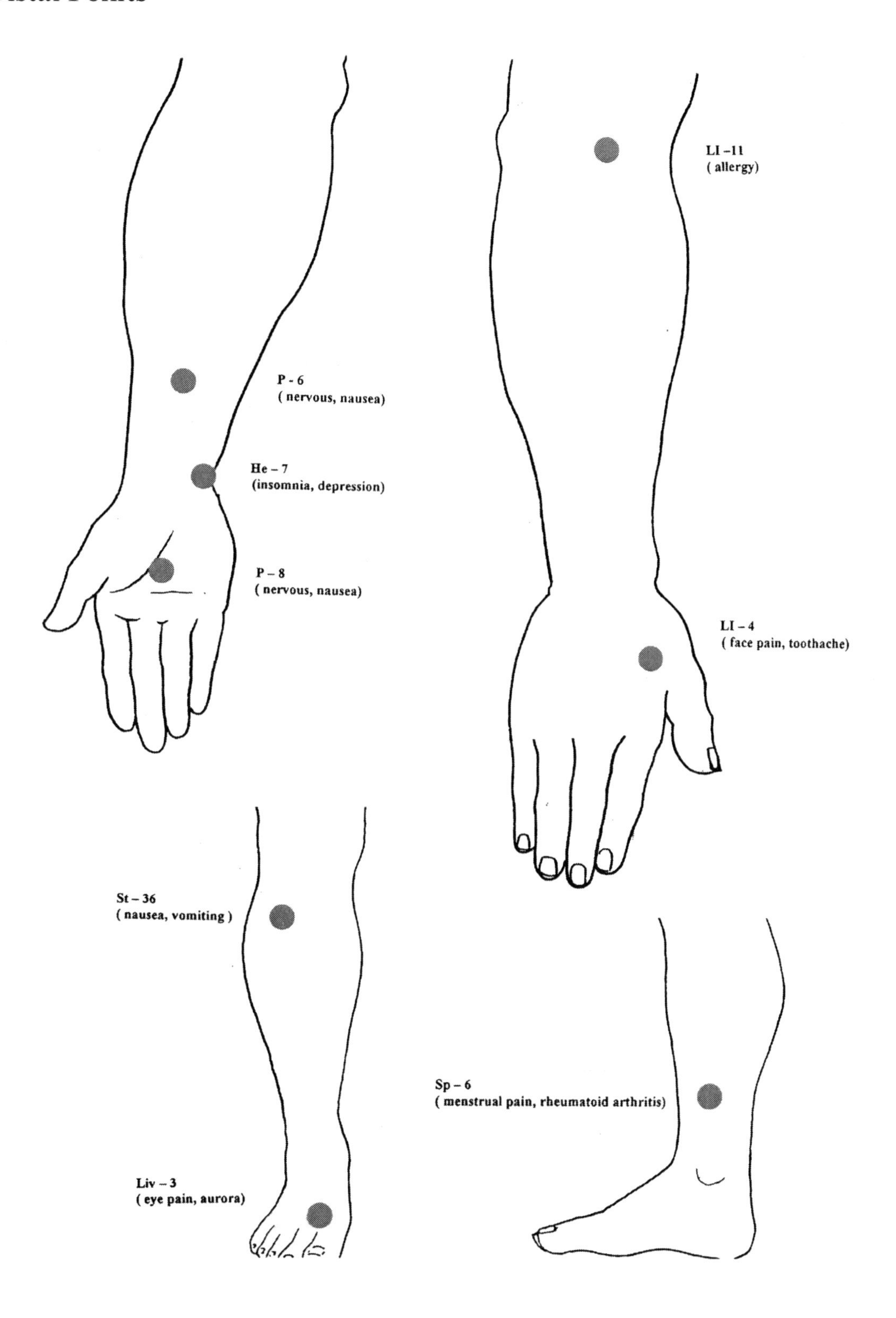

Treatment of Headache
Table 4.
Stimulation Sites on the Local Vascular System

Acupuncture Point (Fig)	Location	Anatomy
GB 26	In the depression between the trapezius and sterno-mastoid below the occipital bone	- Vertebral artery at C. 1 level (vertebral-basilar)
B1 2	Medial upper corner of the obit	Ophthalmic artery (internal carotid)
Tai Yang	In the temporal fossa	Temporal artery (External carotid)

Treatment of Headache
Diagram 3.
Panic Switches (see Table 3)

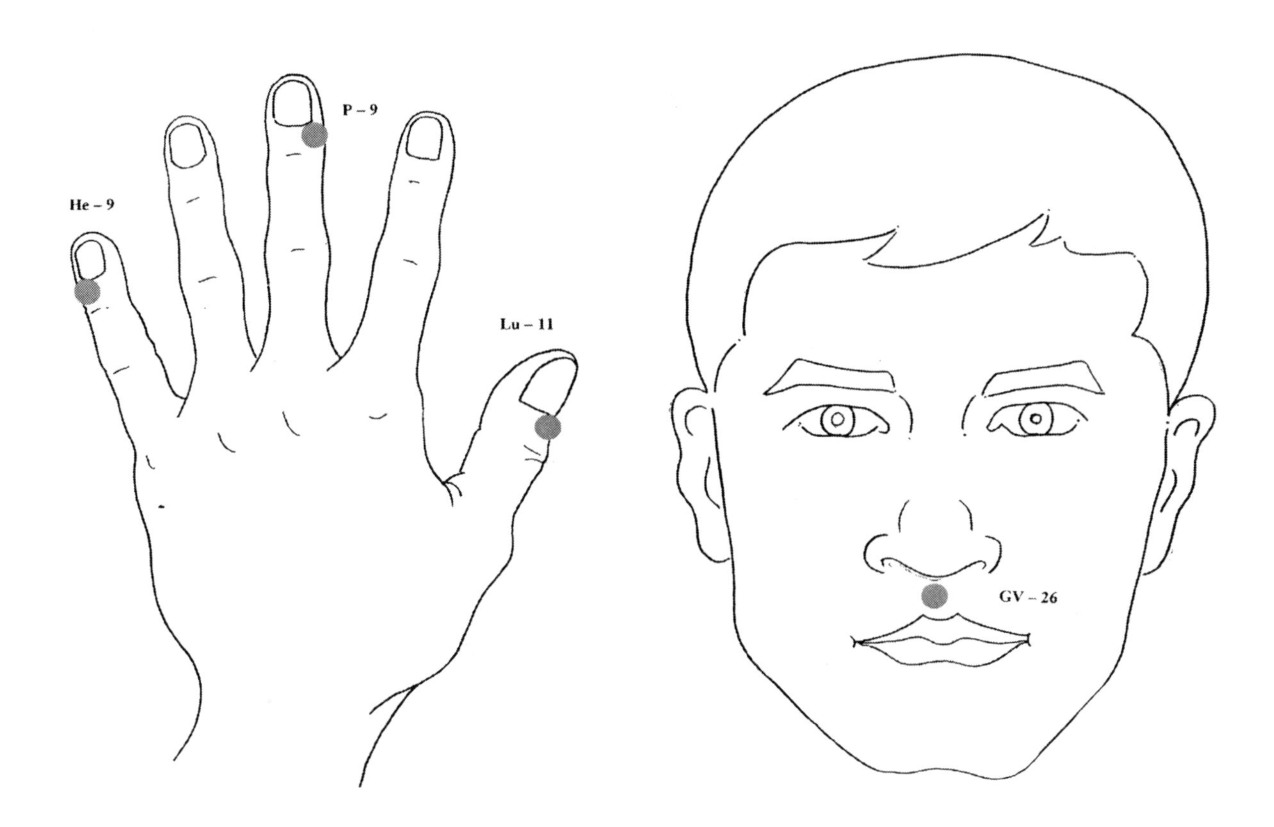

Treatment of Headache
Diagram 4.
Local Vascular Headache Points

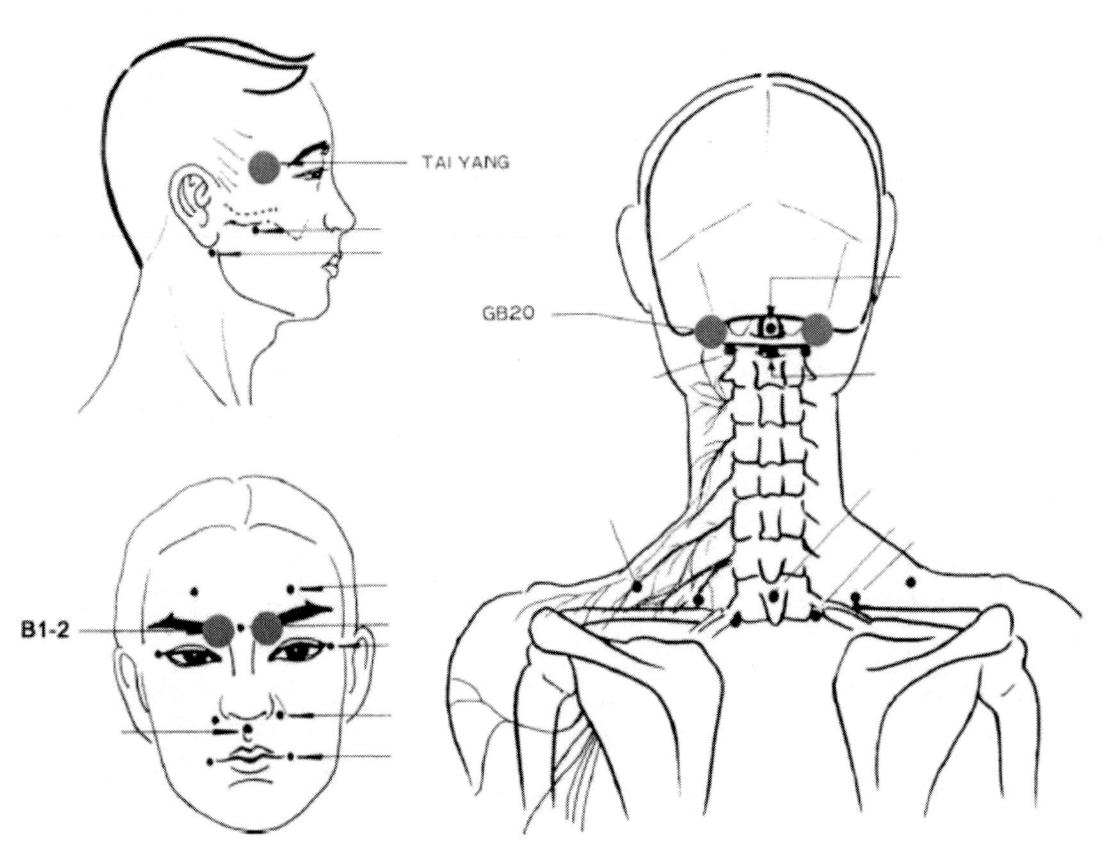

Treatment of Headache
Table 5. Musculotendinous Headache Points

Acupuncture Point (Fig)	Location Innervation	Muscle	Nerve
GB 14	1 inch above the mid-point of the upper orbital margin	Frontalis	Facial nerve
GV 24.5	Midpoint between eyebrows	Procerus	Facial nerve
B1 8	In the posterior quadrant of the scalp - 1.5" from mid-line	Occipitalis	Facial nerve
TW 17	Between the mastoid-styloid tip and the mandible		Facial nerve trunk

Treatment of Headache
Diagram 5.
Musculotendinous Headache Points

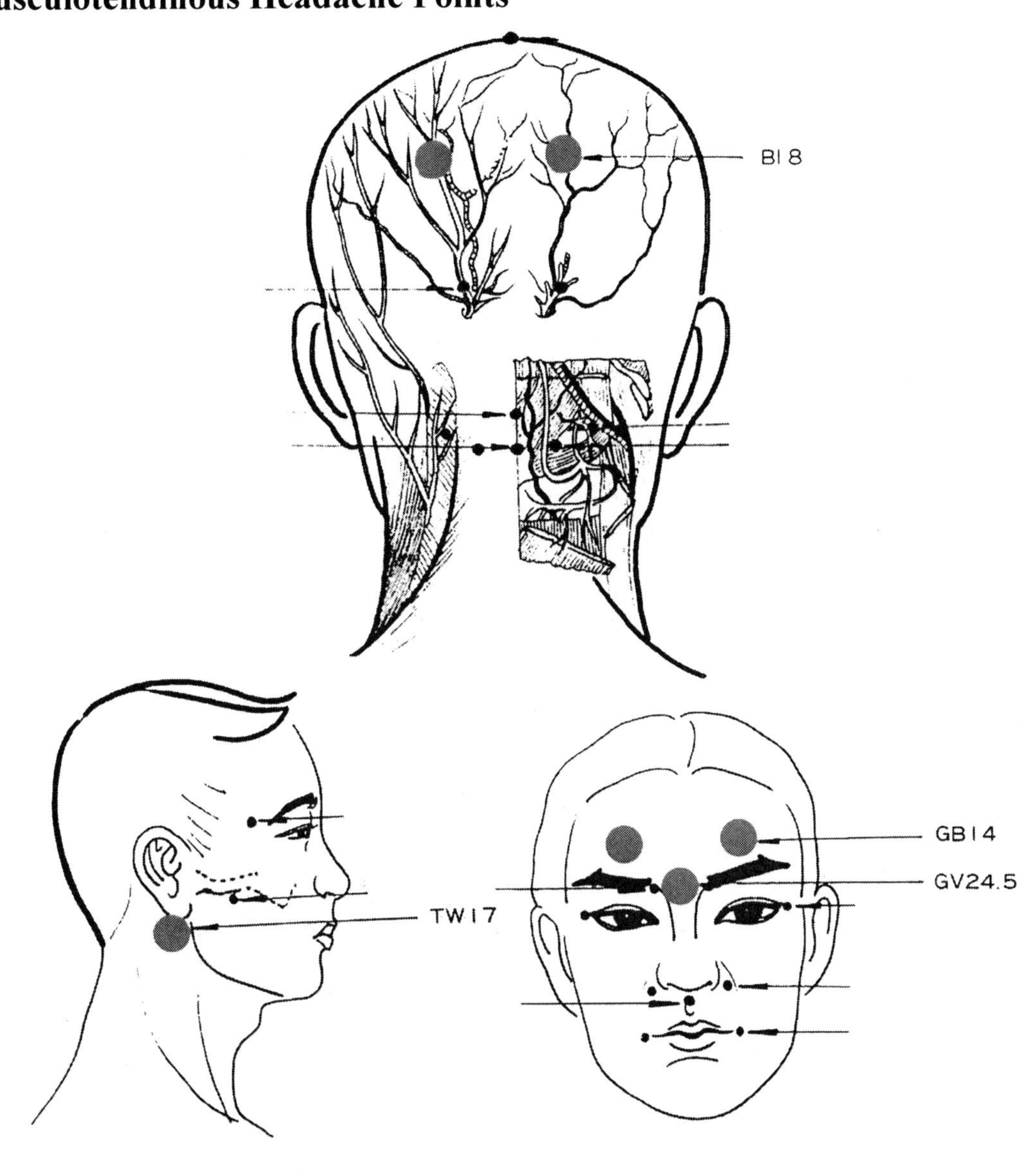

Treatment of Headache
Table 6.
Neuralgic Headache Points

Acupuncture Point (Fig)	Location	Nerve
GV 15	Midpoint above C.2 spinous process	C.2, C.3
B1 9	Lateral to occipital protuberance	
B1 2	Medial upper corner of the orbit	Supratrochlear nerve (V)
LI 20	In the nasolabial groove	Infraorbital (V)
St 7	Between zygomatic arch and mandibular notch	Inf. Alveolar nerve (V)
GB 14	1" above the midpoint of the upper orbital margin.	Temporal nerve
GB 1	0.5" lateral to the external canthus.	Zygomatic nerve
St 4	Angle of the mouth	Facial-cervical nerve
TW 17	Between the mastoid-styloid tip and the mandible	Facial nerve trunk

Treatment of Headache
Diagram 6.
Neuralgic Headache Points

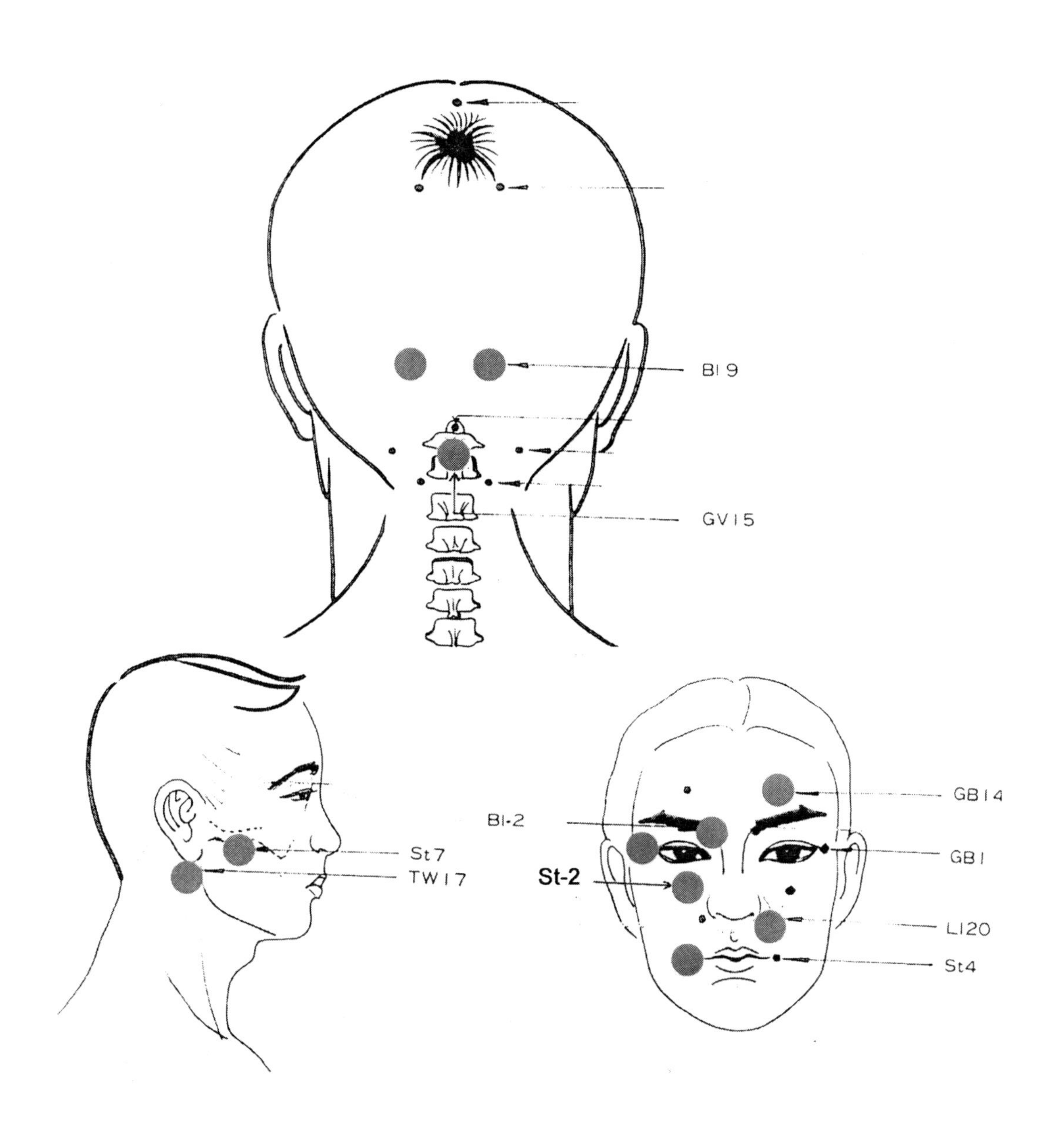

Musculoskeletal Disorders of the Neck
2 a) Table 7.
Cervical Nerve Roots and Symptoms

Spinal Segment	Radicular Symptoms
C.3	Pain and numbness in the posterior cervical area around mastoid process and pinna of the ear
C.4	Pain and numbness in the posterior cervical area radiating along levator scapula and occasionally to anterior chest
C.5	Pain radiating from the lateral cervical area to shoulder top; numbness over middle of deltoid
C.6	Pain radiating down lateral side of arm and forearm often to thumb and index finger
C.7	Pain radiating to middle of forearm and to middle three fingers of hand
C.8	Pain radiating to forearm and to ring and little finger with Numbness rarely extends above the wrist

Musculoskeletal Disorders of the Neck
2 a) Table 8.
Acupuncture Points in the Neck

Acupuncture Point (Fig)	Location	Nerve
GV 16	Midpoint between the occipital bone and C.1 vertebrae	C.1 nerve root
GV 15	Midpoint between the C.1 and C.2 vertebrae	C.2 nerve root
GV 14	Midpoint between the C.7 and T.1 vertebrae	C.8 and T.1
Bl 10	Just below the 1st cervical vertebrae at the level of its posterior tubercle	C.2 nerve root
Bl 11	1.5" from the mid-spine line at the level of the lower margin of C.7 vertebrae	C.8 and T.1 nerve root
GB 20	In the fossa between the trapezius and sterno-mastoid	Great occipital nerve vertebrae artery
GB 21	Halfway between the cervical prominence of C.7 and the acromion	Trunk of accessory nerve
Sl 15	2" lateral to GV 14	Dorsal scapular nerve
Sl 17	Between the mandible angle and the sterno-mastoid	Superior cervical sympathetic ganglion
St 9	1.5" lateral to the thyroid cartilage	Carotid sinus sympathetic trunk

Musculoskeletal Disorders of the Neck
2a) Diagram 7.

Spastic Torticollis

Cervical Spine:

Spasmodic Torticollis

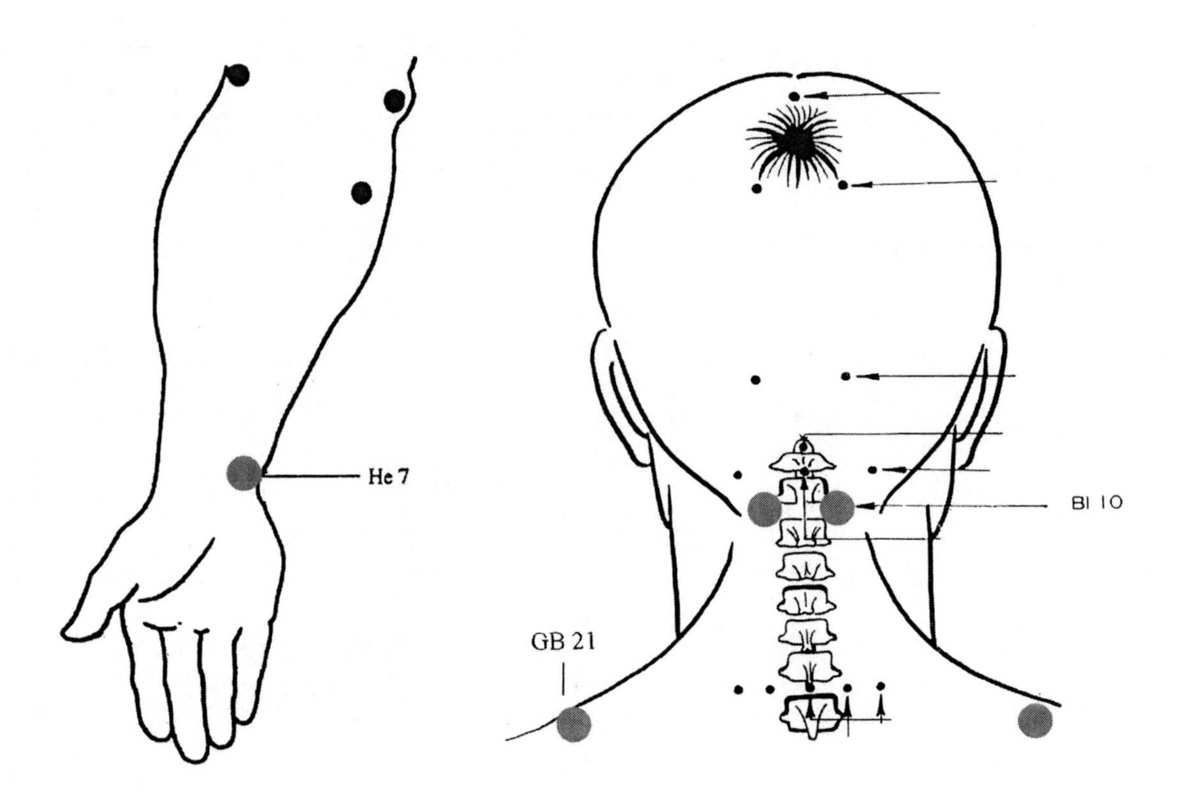

Musculoskeletal Disorders of the Neck
2a) Diagram 8.
Cervical Spine: Fibrositis

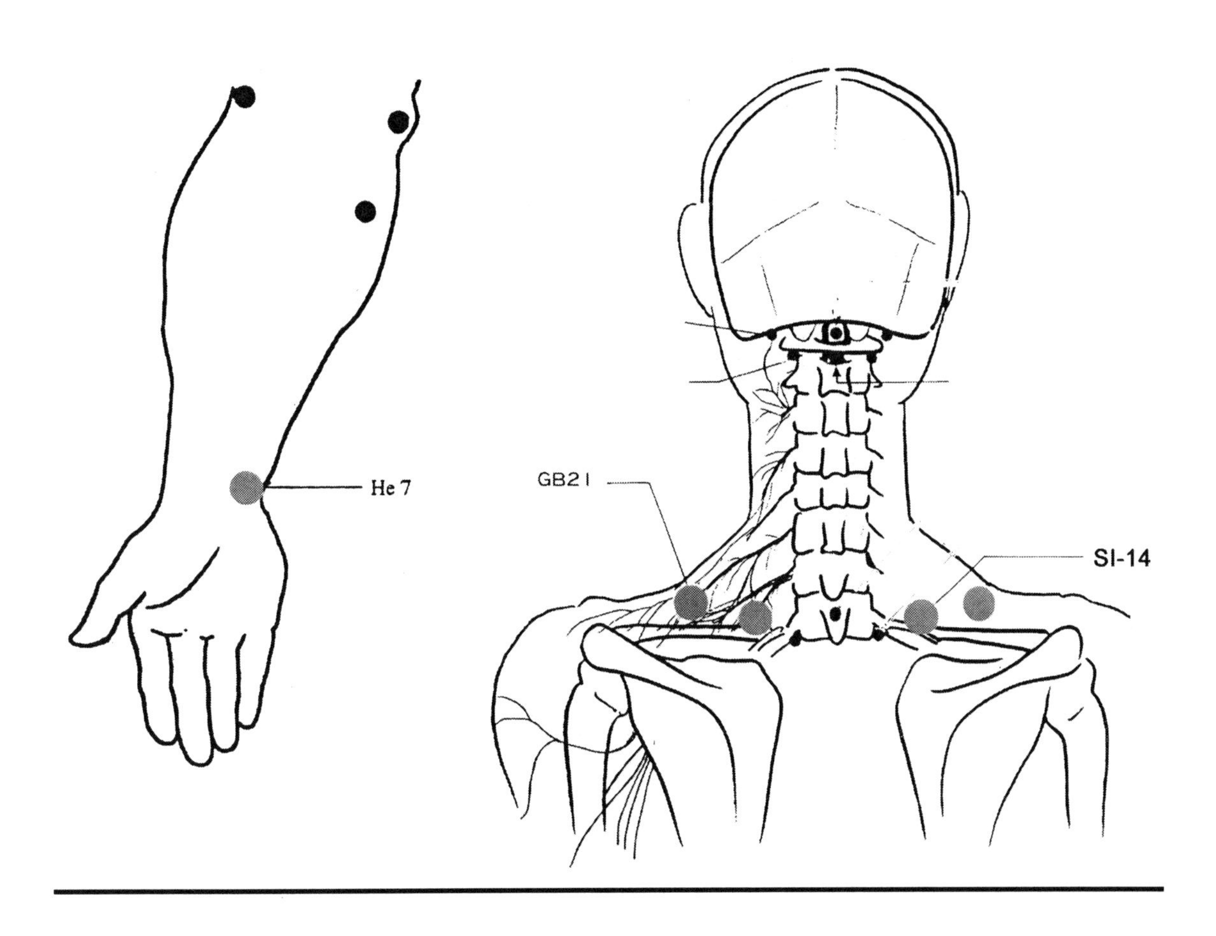

Musculoskeletal Disorders of the Neck
2a) Diagram 9.
Cervical Spine: Whiplash

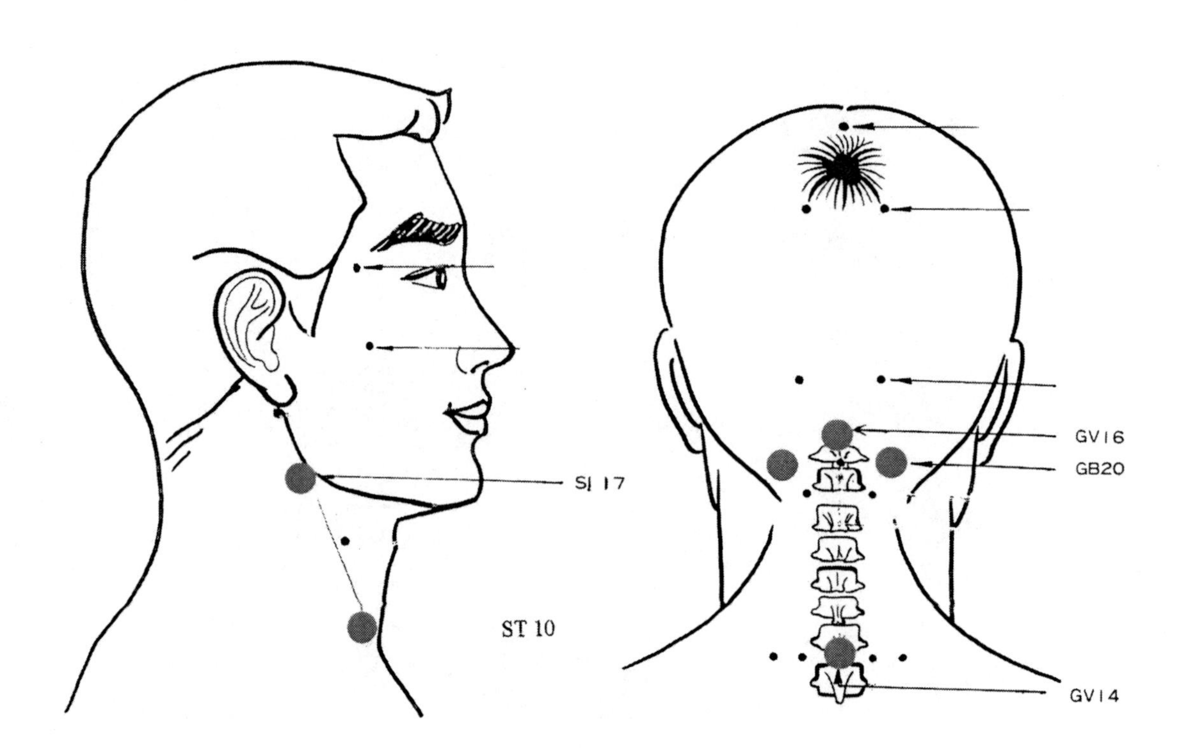

Musculoskeletal Disorders of the Shoulder
2 b) Table 9.
Acupuncture Points for Disorders of the Shoulder

Spinal Segment	Disorder	Acupuncture Point (Fig)	Location	Anatomy
C.4-6	Supraspinatus tendinitis	LI 15	Greater tubercle of humerus	Tendon insertion
		LI 16	Behind the acromio-clavicular arch	Musculo-tendinous junction
		SI 12	Mid point in supraspinatus fossa	Suprascapular nerve
C.5-6	Sub-deltoid bursitis	TW 14	Neck of humerous	Circumflex nerve
		LI 14	Above the deltoid tubercle of humerus	Tendon insertion
C.5-6	Infraspinatus tendinitis	SI 11	Midpoint in infraspinatus fossa	Suprascapular nerve
C.5-6	Bicipital tendinitis	LI 15 Lu 3	Between humerus and biceps muscles (3" below axilla)	Musculo-cutaneous nerve
C.5-6	Adhesive capsulitis	Bicipital groove point TW 14 LI 15	Bicipital groove	Tendon insertion

Musculoskeletal Disorders of the Shoulder
2b) Diagram 10.
Supraspinatus Tendinitis

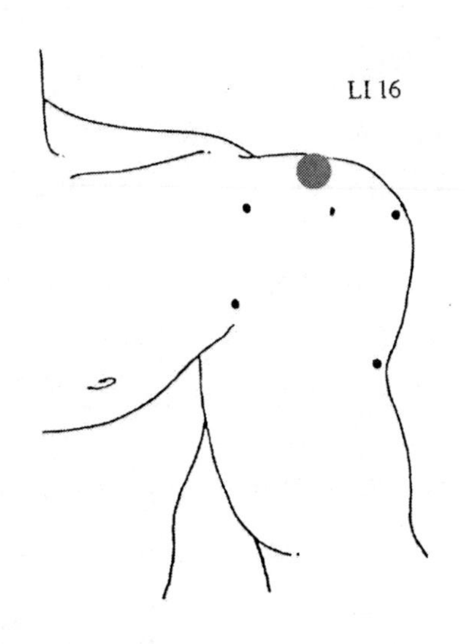

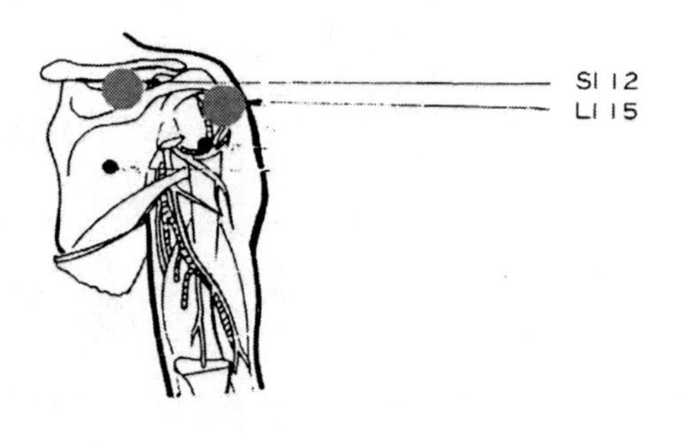

Infra-Spinatus Tendinitis

Sub-Detloid Bursitis

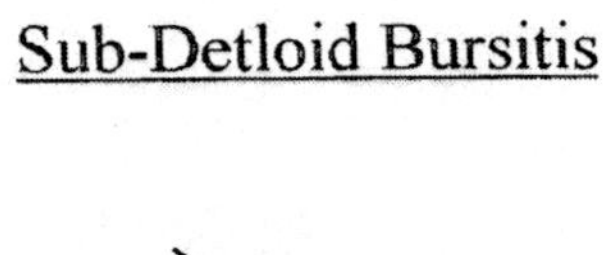

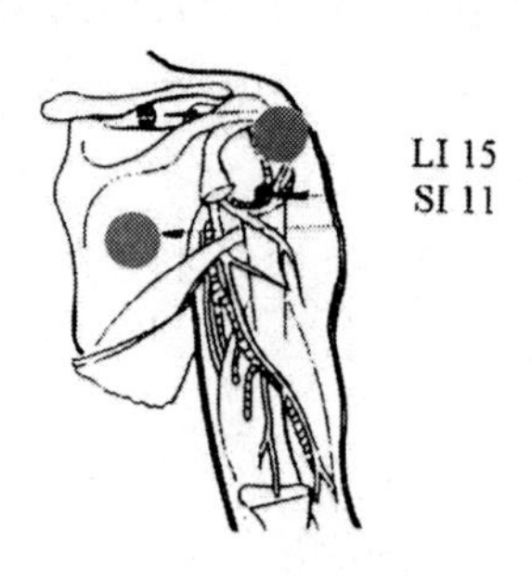

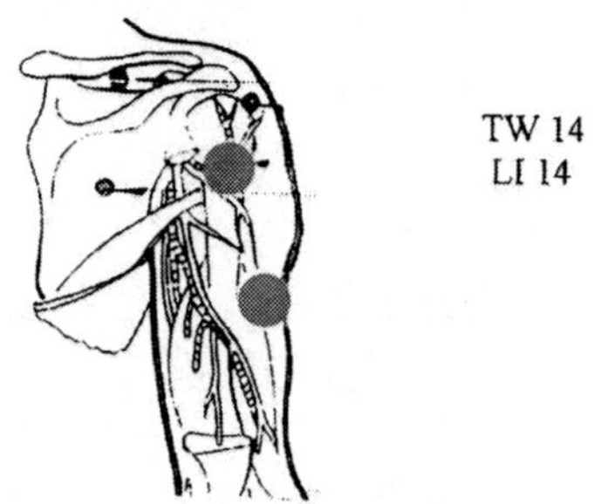

Adhesive Capsulitis

Bicipital Tendinitis

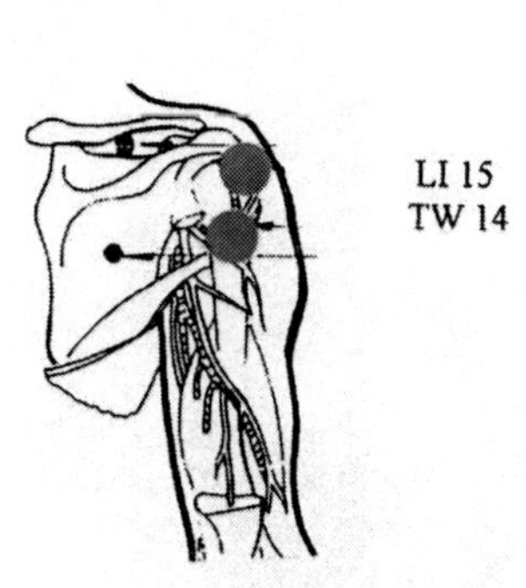

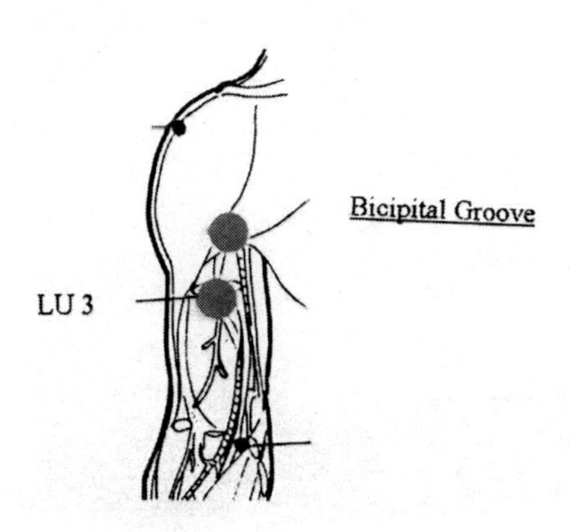

Musculoskeletal Disorders of the Shoulder
2b) Diagram 11.
Shoulder Pain

a) LATERAL SIDE

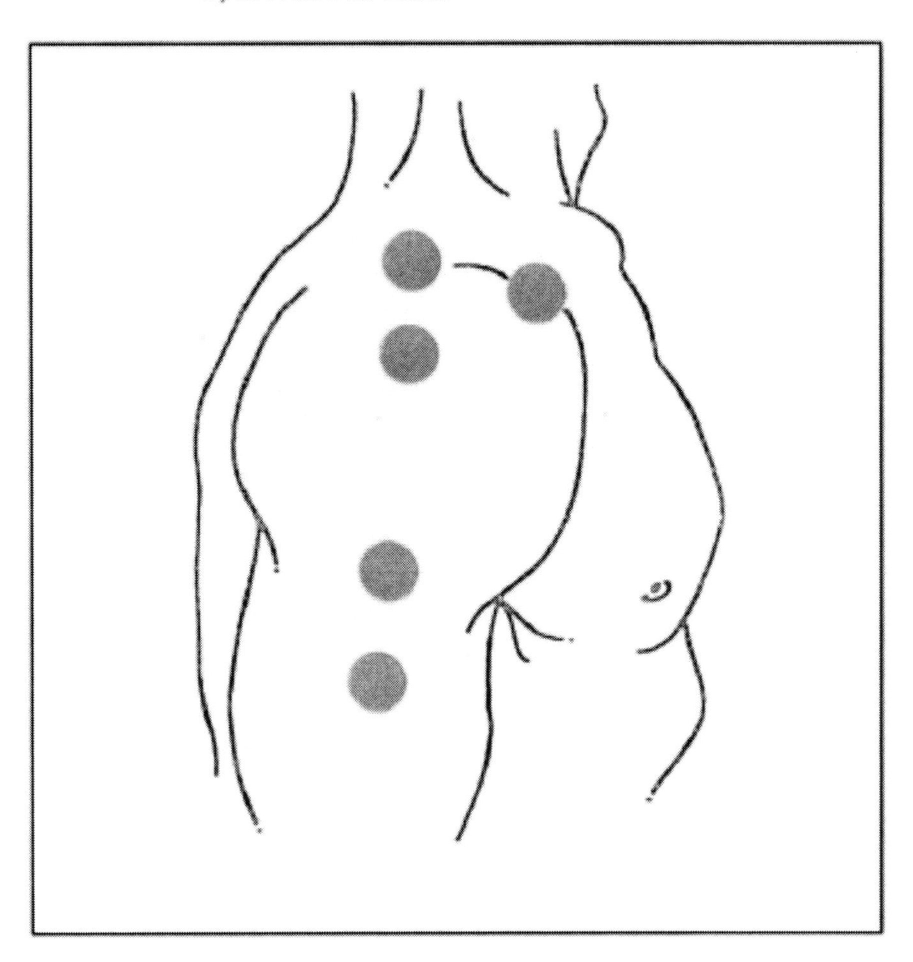

b) ANTERIOR SIDE

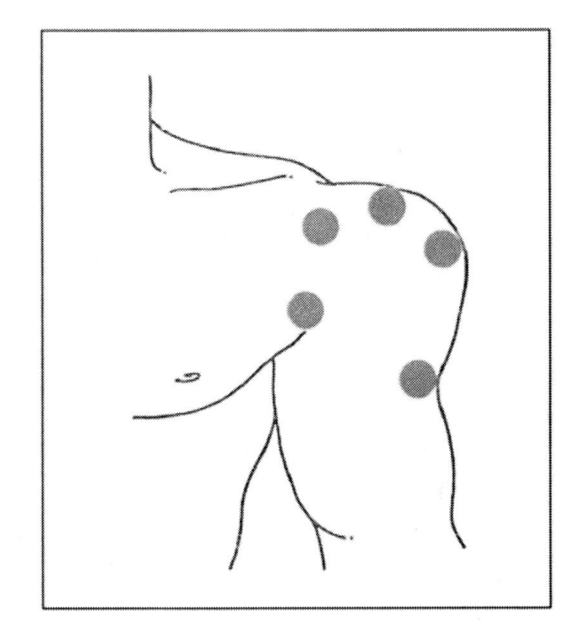

c) POSTERIOR SIDE

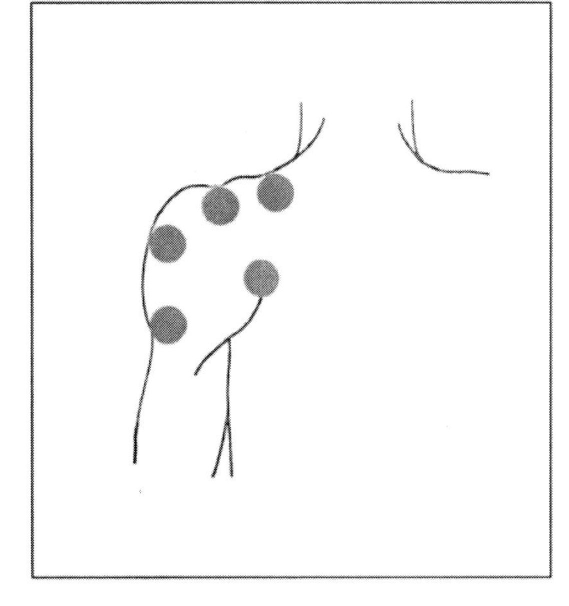

Shoulder Tendinitis

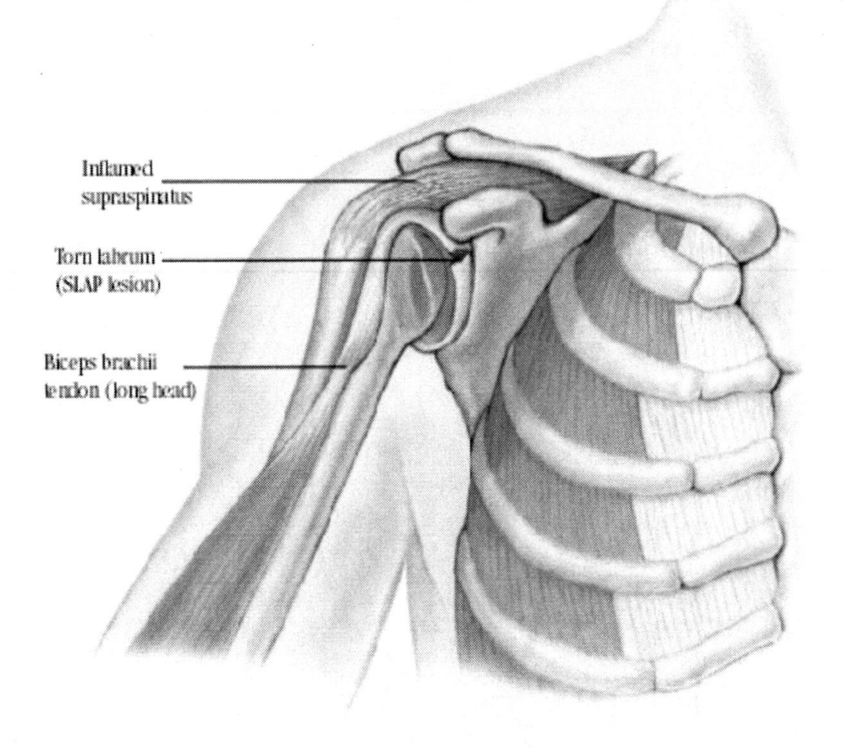

Shoulder Injuries

Dislocated and Separated Shoulder

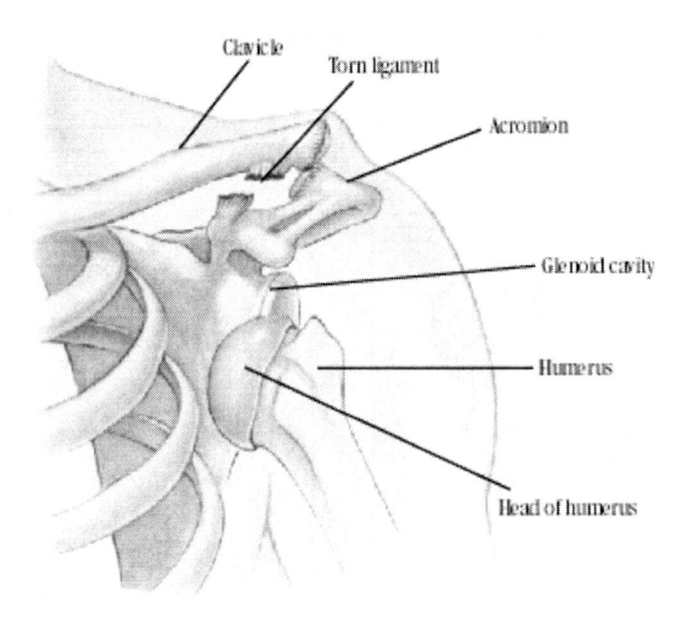

Musculoskeletal Disorders of the Elbow
2 c) Table 10.
Acupuncture Points for Disorders of the Elbow

Spinal Segment	Disorder	Acupuncture Point (Fig)	Location	Anatomy
C.6-8	Lateral Epicondylitis (tennis elbow)	LI 11	Midpoint on elbow crease between lateral epicondyle and biceps tendon	Radial nerve
		LI 10	2" below LI 11	Extensor carpi radialis
C.7-8	Medial Epicondylitis (golfer's elbow)	P 3	Medial to biceps tendon on elbow crease	Median nerve
		SI 8	Mid-point between medial epicondyle and olecranon	Ulnar nerve

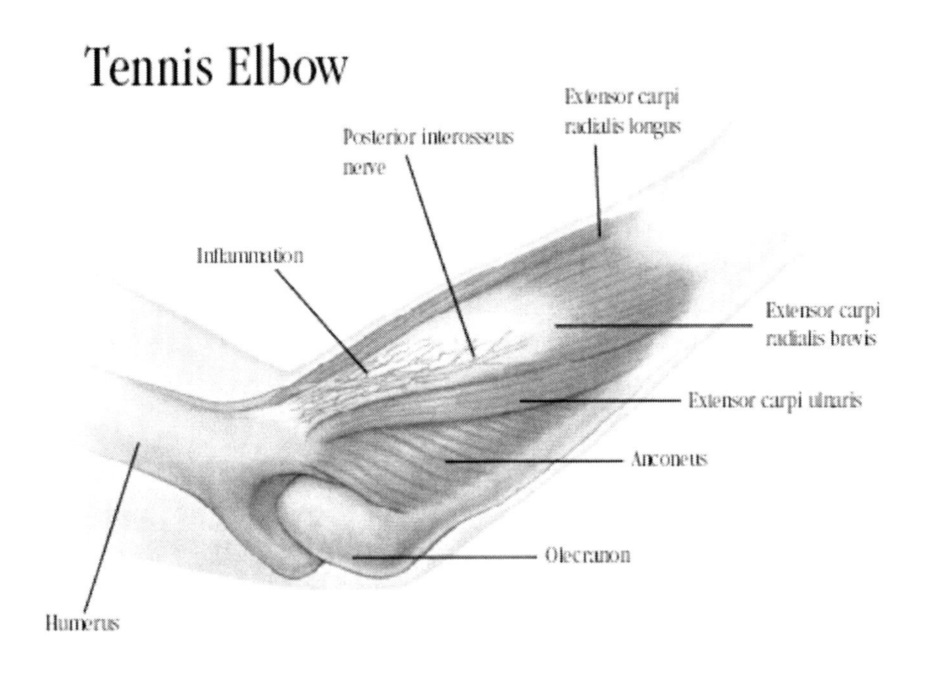

Musculoskeletal Disorders of the Elbow
2c) Diagram 12.
Tennis Elbow (Lateral Epicondylitis)

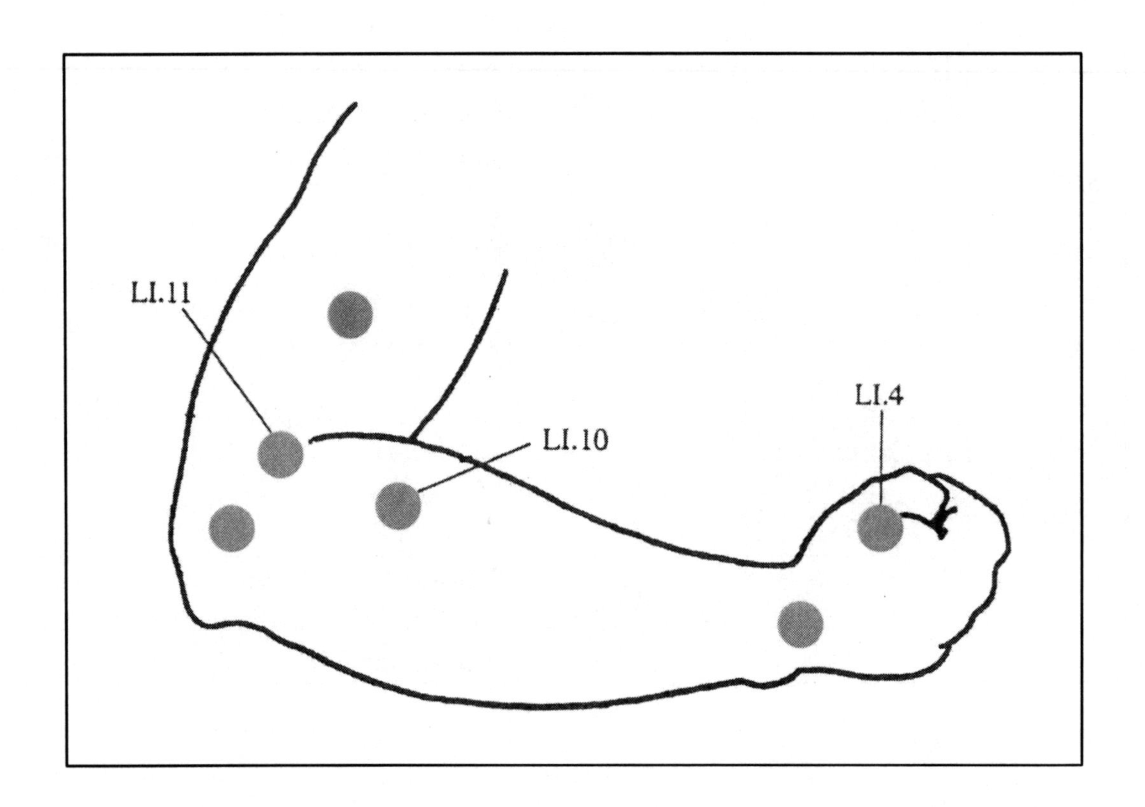

Musculoskeletal Disorders of the Elbow
2c) Diagram 13.
Golfer's Elbow (Medial Epicondylitis)

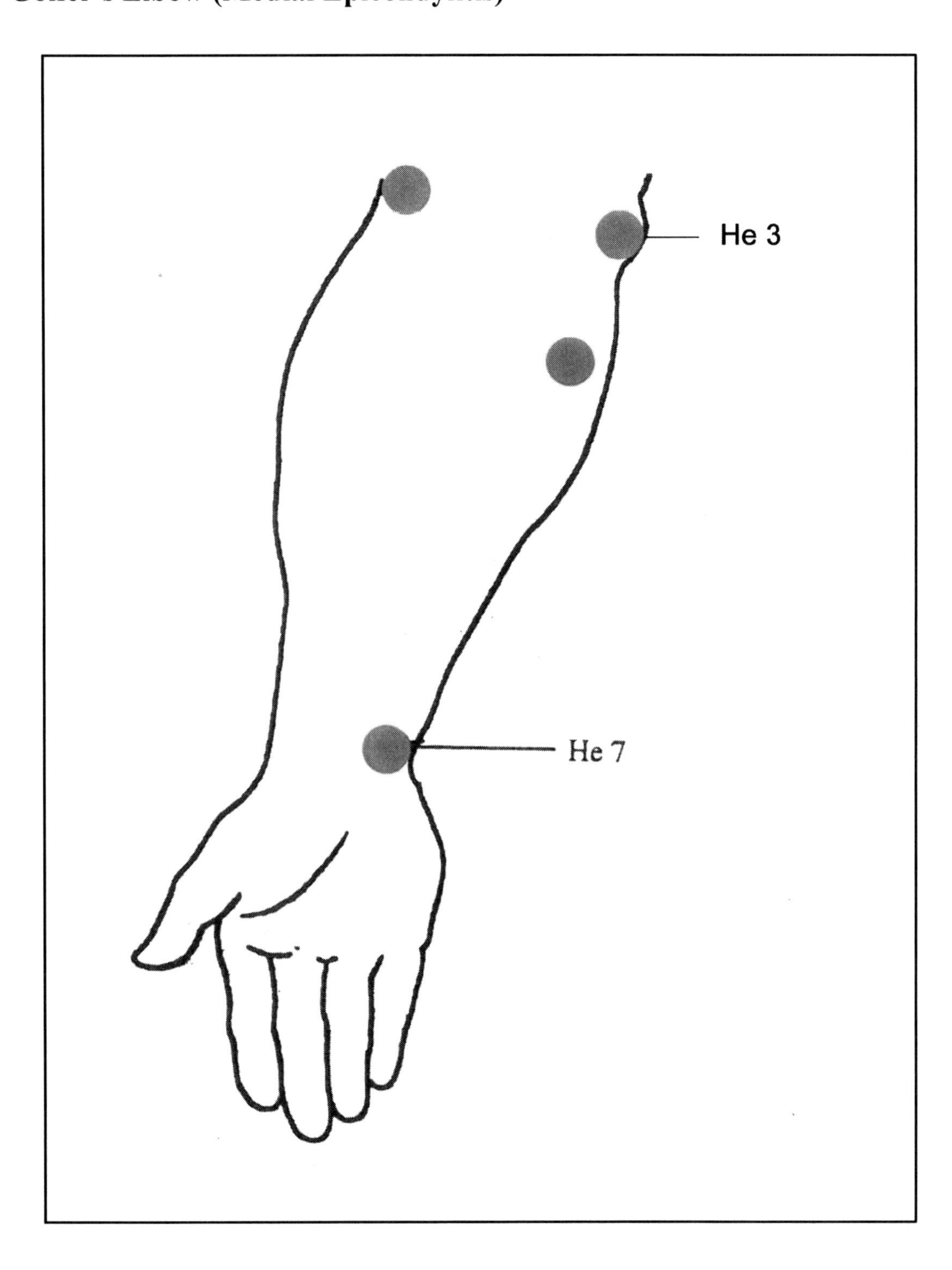

Musculoskeletal Disorders of the Forearms, Wrist, Hands & Fingers
2d) Diagram 14.
Carpal Tunnel Pain

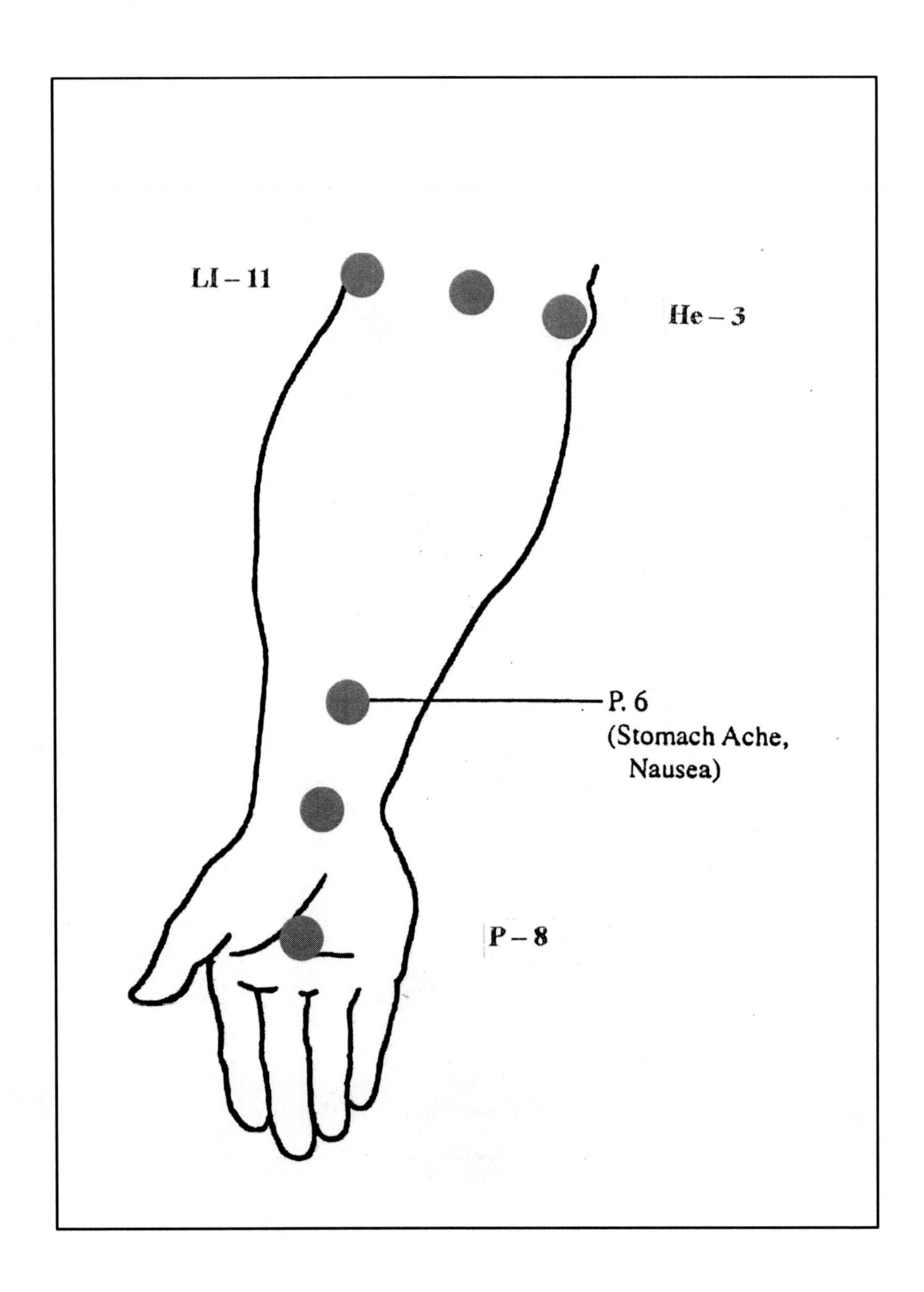

Musculoskeletal Disorders of the Forearms, Wrist, Hands & Fingers
2d) Diagram 15.
Wrist Pain: a) Teno-Synovitis

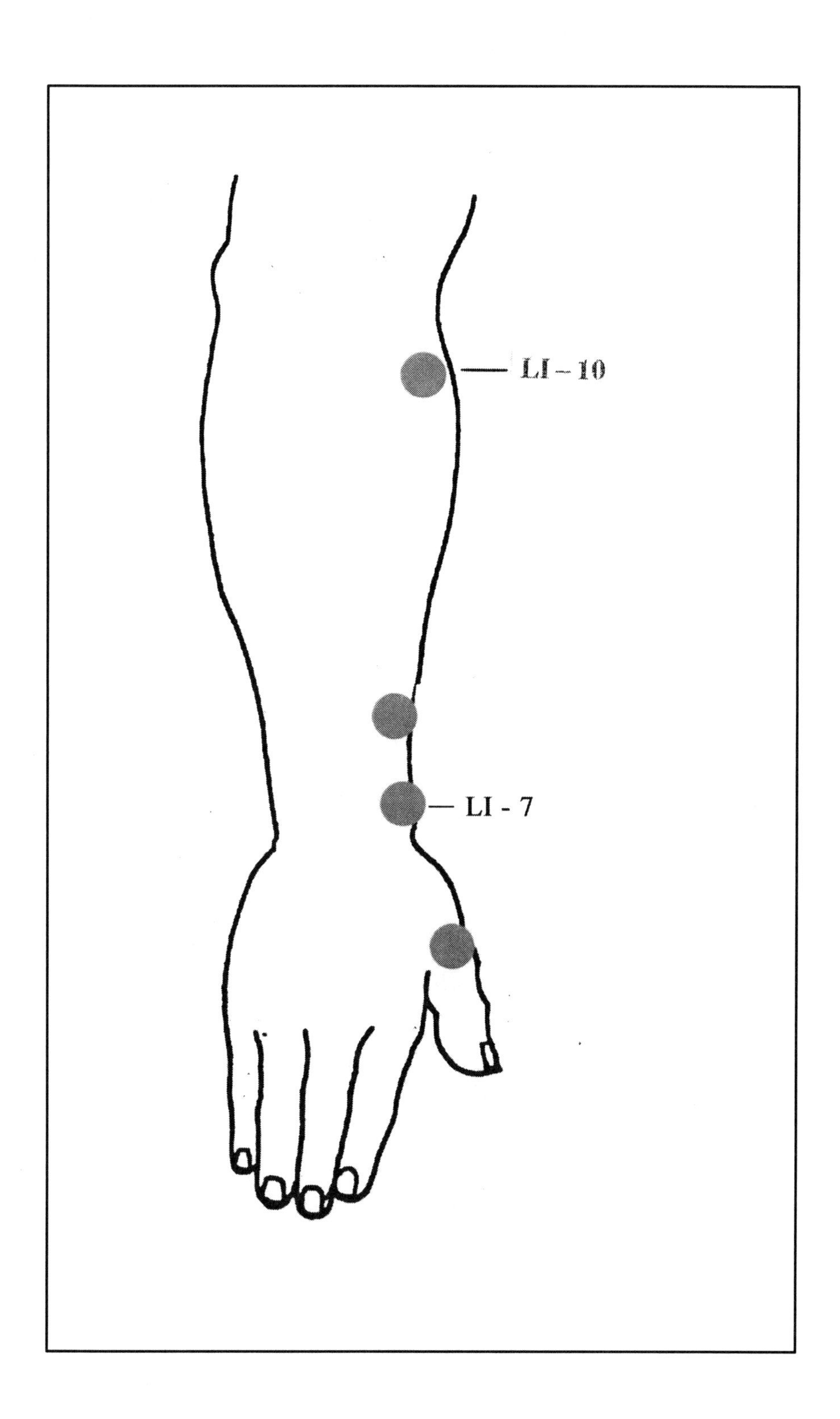

Musculoskeletal Disorders of the Forearms, Wrist, Hands & Fingers
2d) Diagram 16.
Arm and Hand Pain (Ventral)

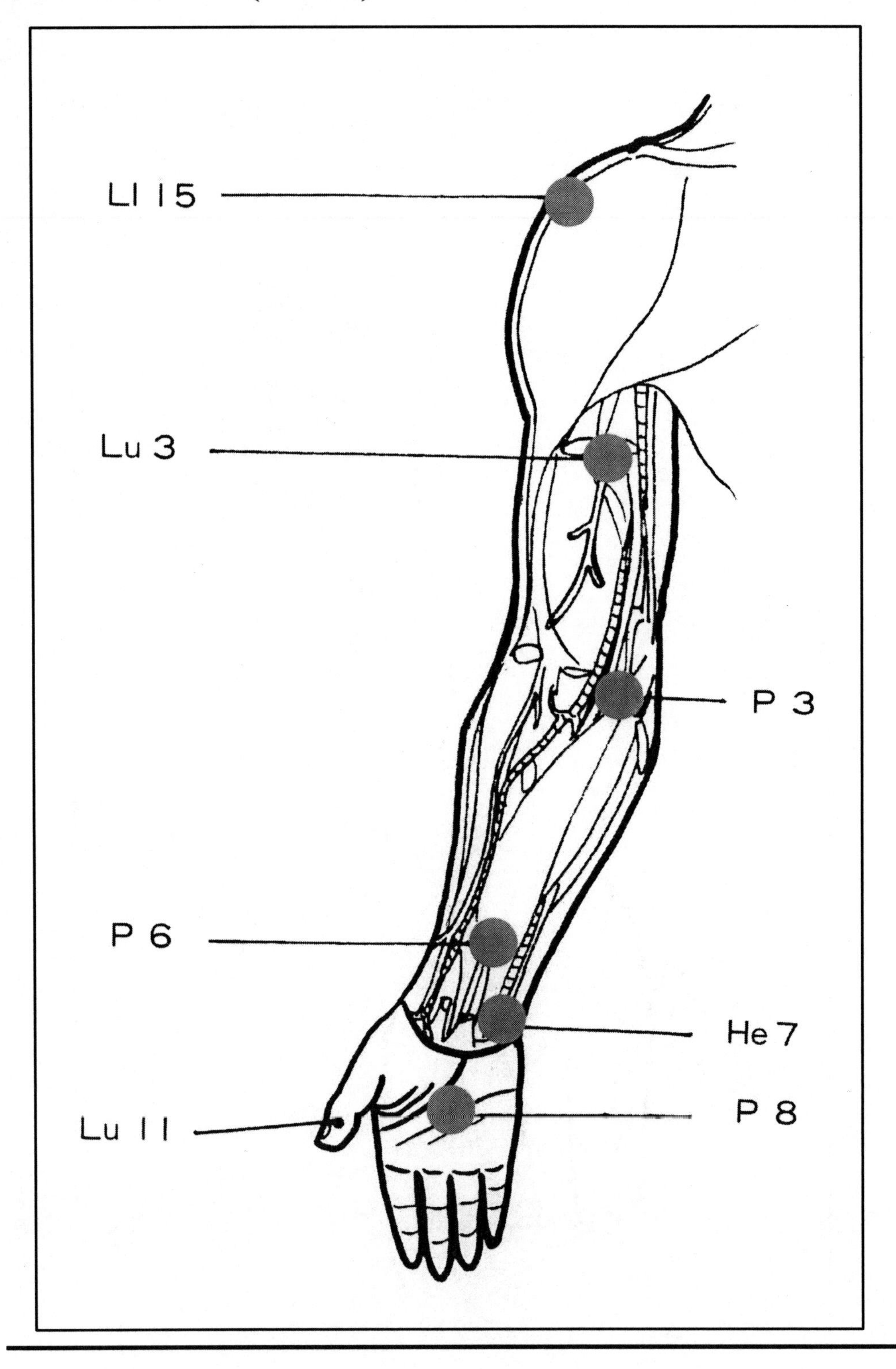

Musculoskeletal Disorders of the Forearms, Wrist, Hands & Fingers
2 d) Table 11.
Acupuncture Points for Disorders of the Wrist

Spinal Segment	Disorder	Acupuncture Point (Fig)	Location	Anatomy
C.6-8	Traumatic arthritis Osteoarthritis	LI 5	Distal to radius between tendons of EPL and EPB	Radial nerve
C.7, 8 T.1	Ligamentous sprain	TW 4	Distal to ulna between tendons of ED and EDM	Ulnar nerve

EPL = Extensor Pollicis Longus
EPB = Extensor Pollicis Brevis
ED = Extensor Digitorium
EDM = Extensor Digitminimi

Musculoskeletal Disorders of the Forearms, Wrist, Hands & Fingers
2d) Diagram 17.
Arm and Hand Pain (Dorsal)

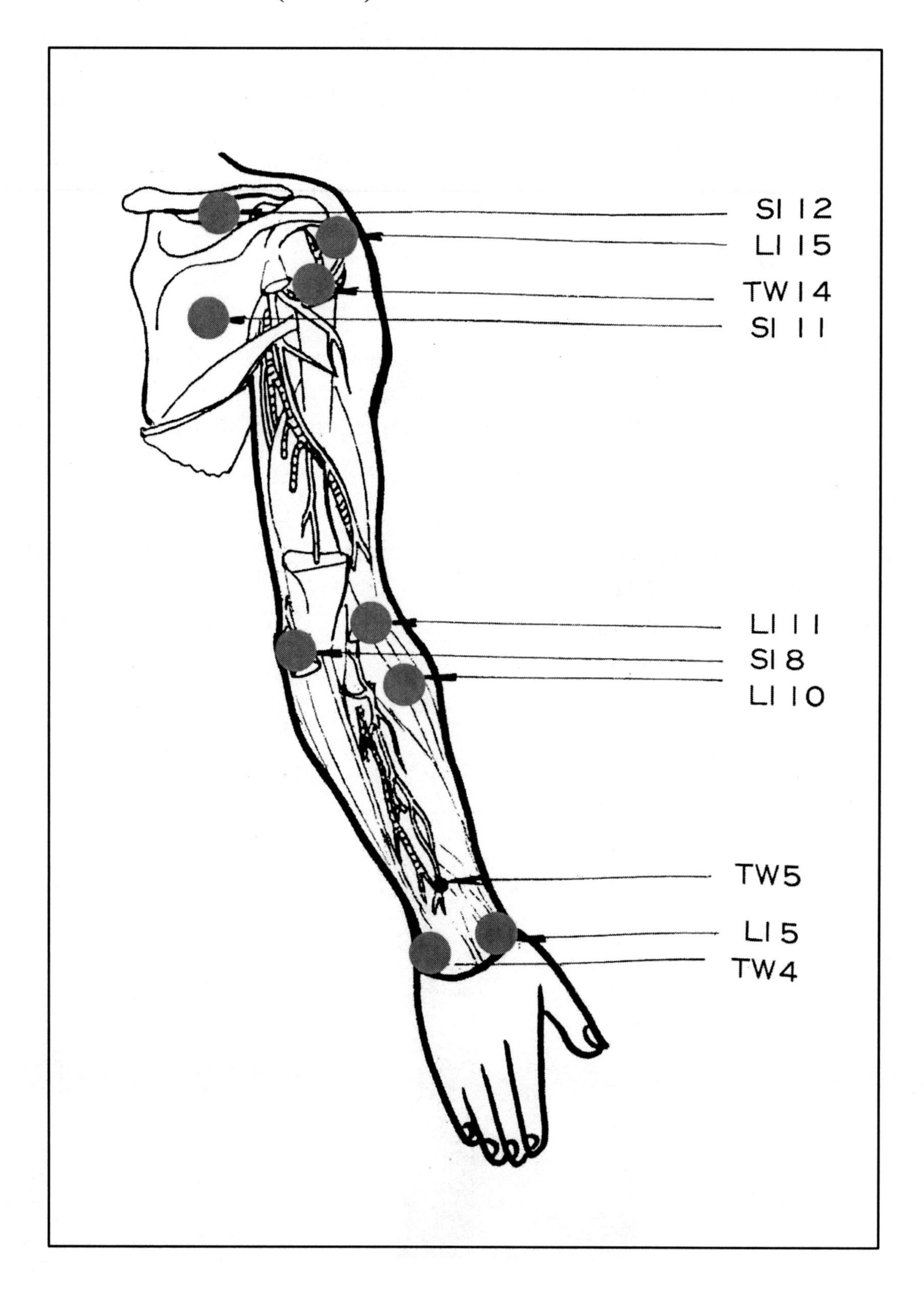

Musculoskeletal Disorders of the Forearms, Wrist, Hands & Fingers
2d) Diagram 18.
Wrist Pain: b) Osteoarthritis, Sprain, Rheumatoid Arthritis

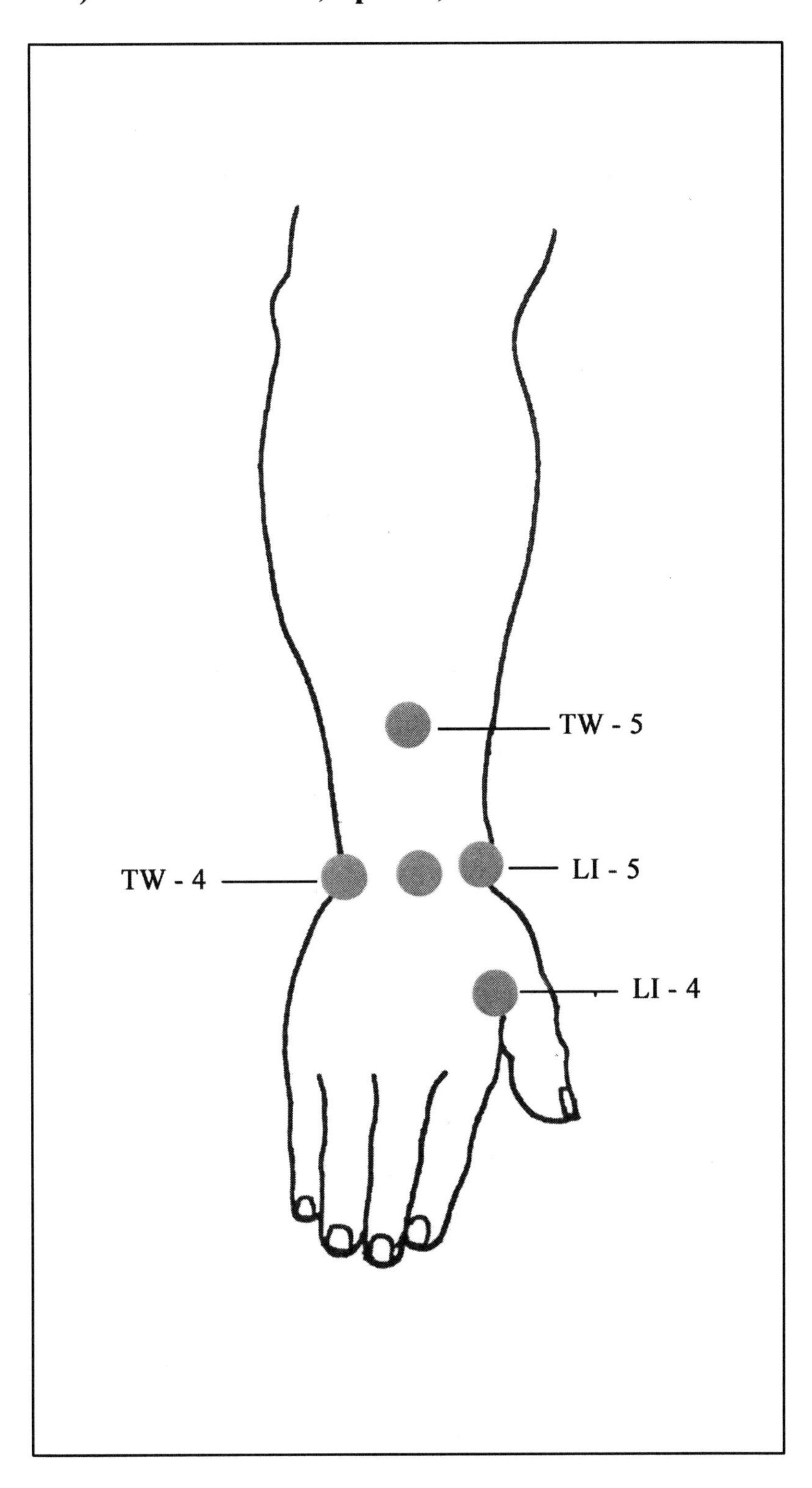

Musculoskeletal Disorders of the Forearms, Wrist, Hands & Fingers
2d) Diagram 19.
Finger Joint Pain:
a) Dorsal Approach, Osteoarthritis, Rheumatoid Arthritis

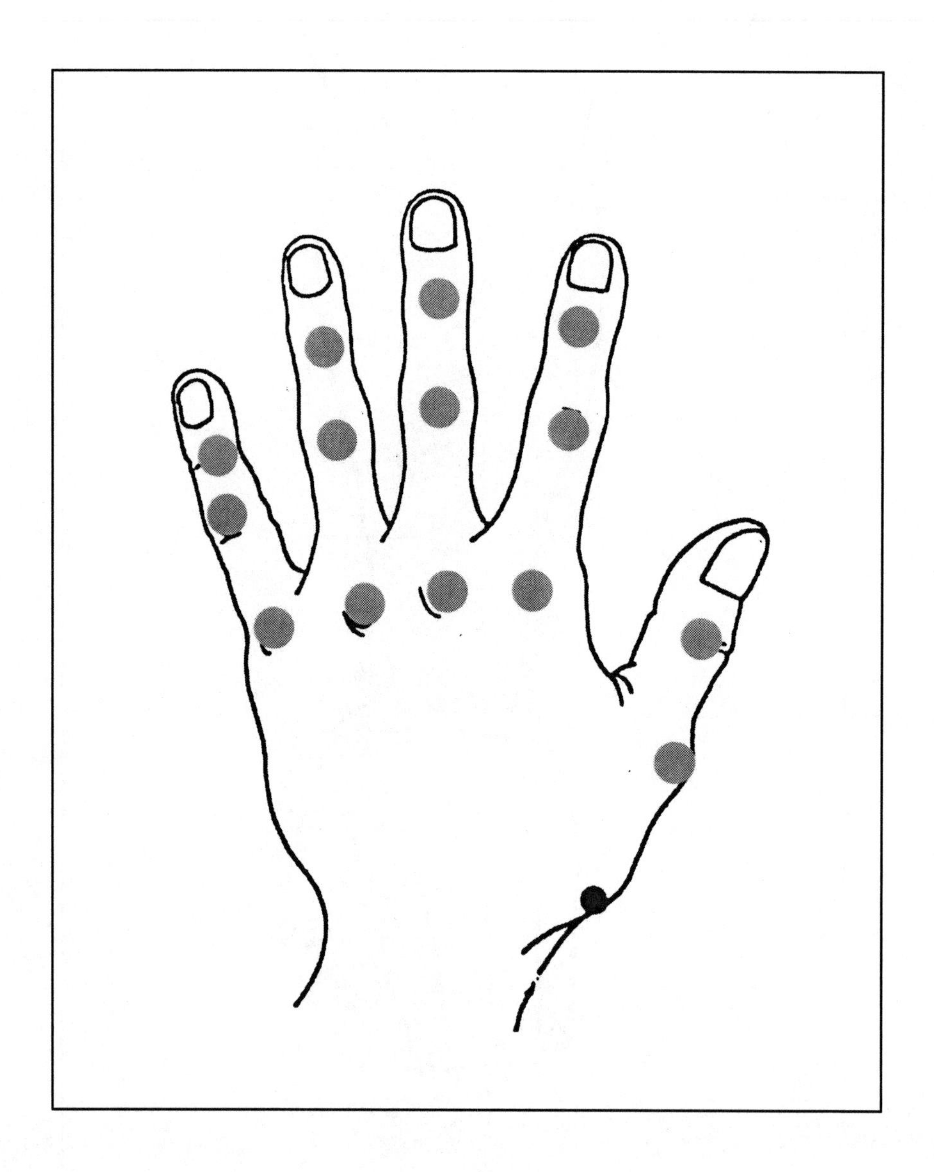

Musculoskeletal Disorders of the Forearms, Wrist, Hands & Fingers
2d) Diagram 20.
Finger Joint Pain:
a) Ventral Approach, Osteoarthritis, Rheumatoid Arthritis

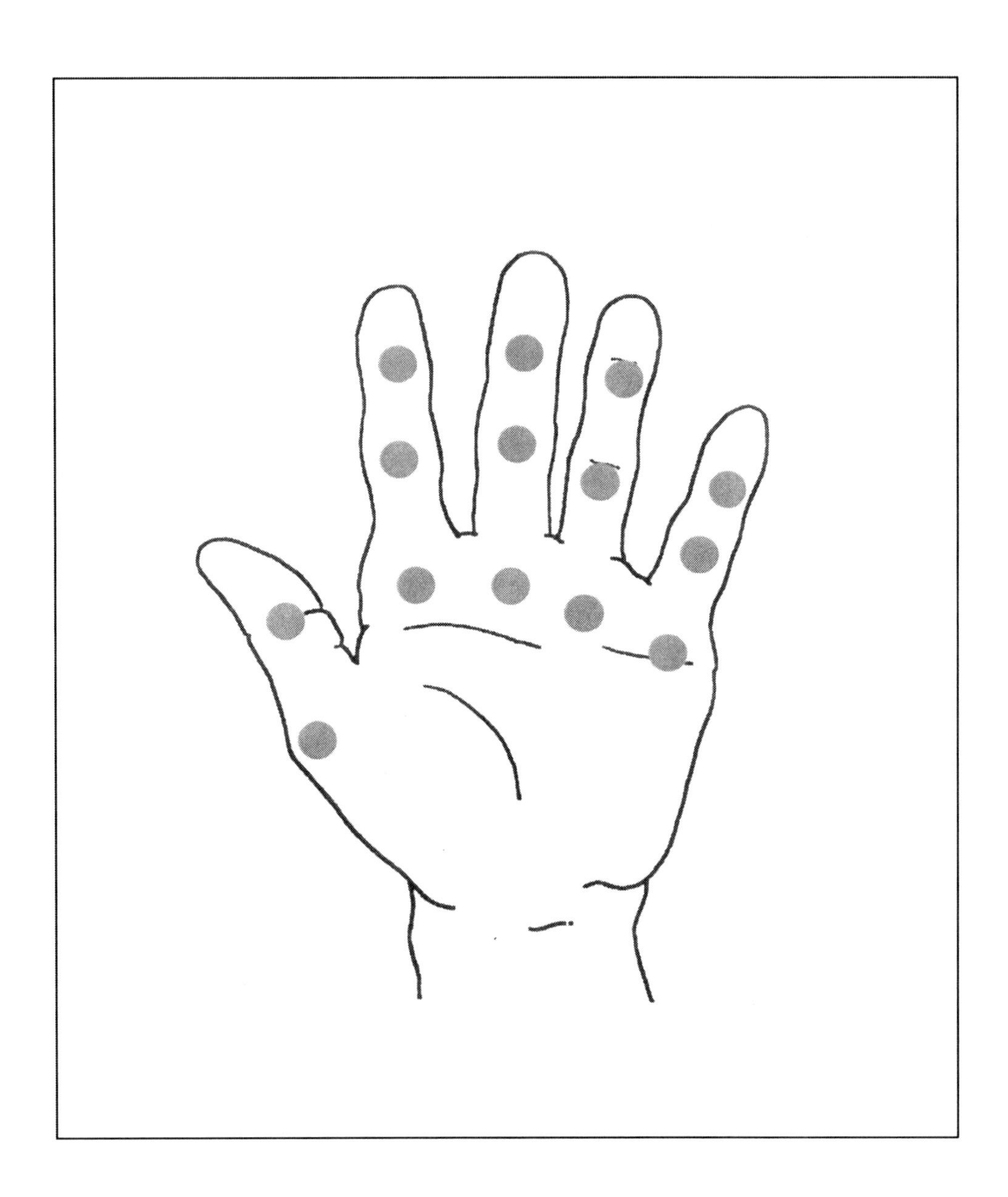

Musculoskeletal Disorders of the Upper Back
2e) Diagram 21.
Upper Back Pain (Muscular Pain)

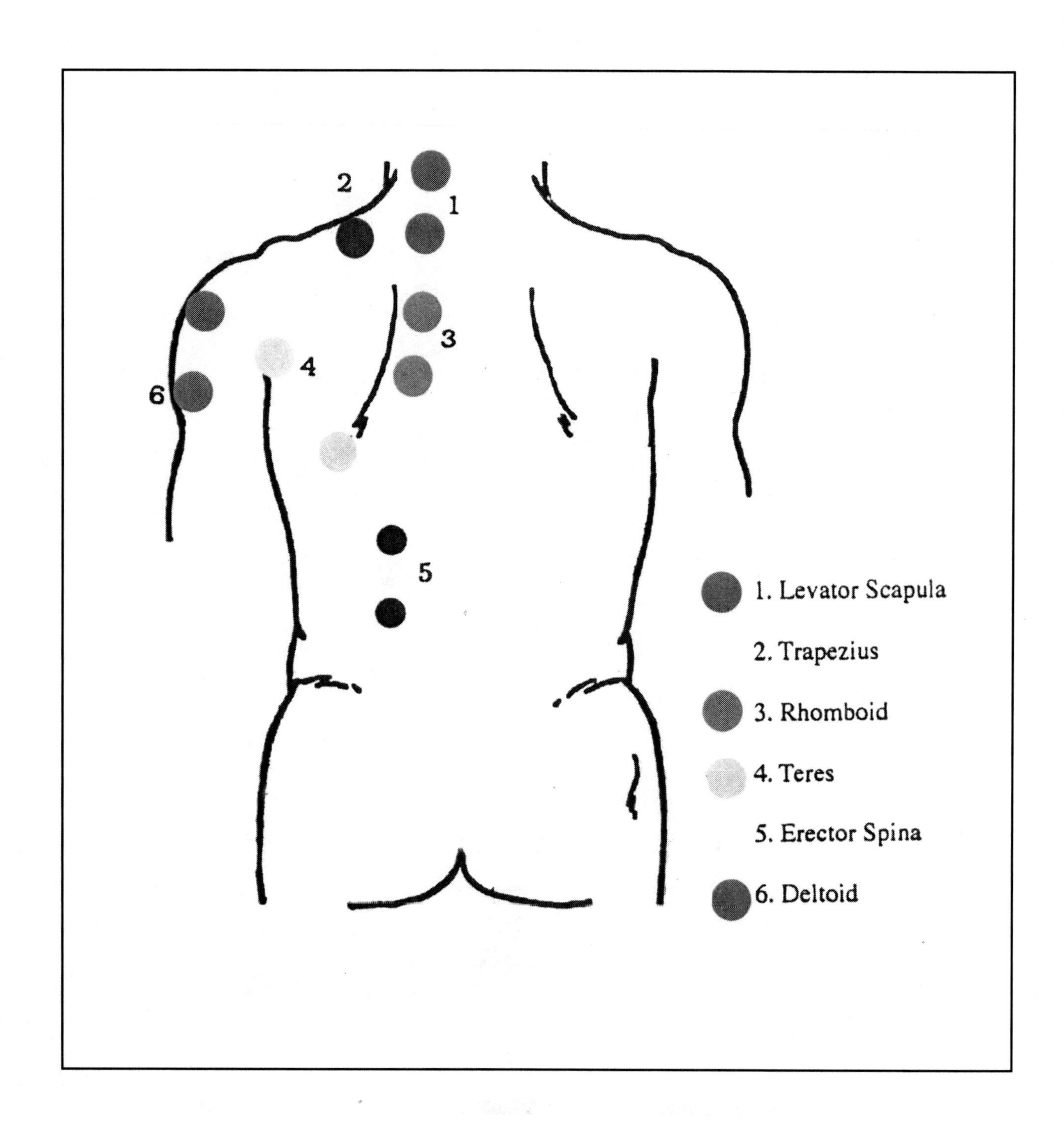

Musculoskeletal Disorders of the Chest and Abdomen
2f) Diagram 22.
Chest Pain (Emphysema, Asthma)

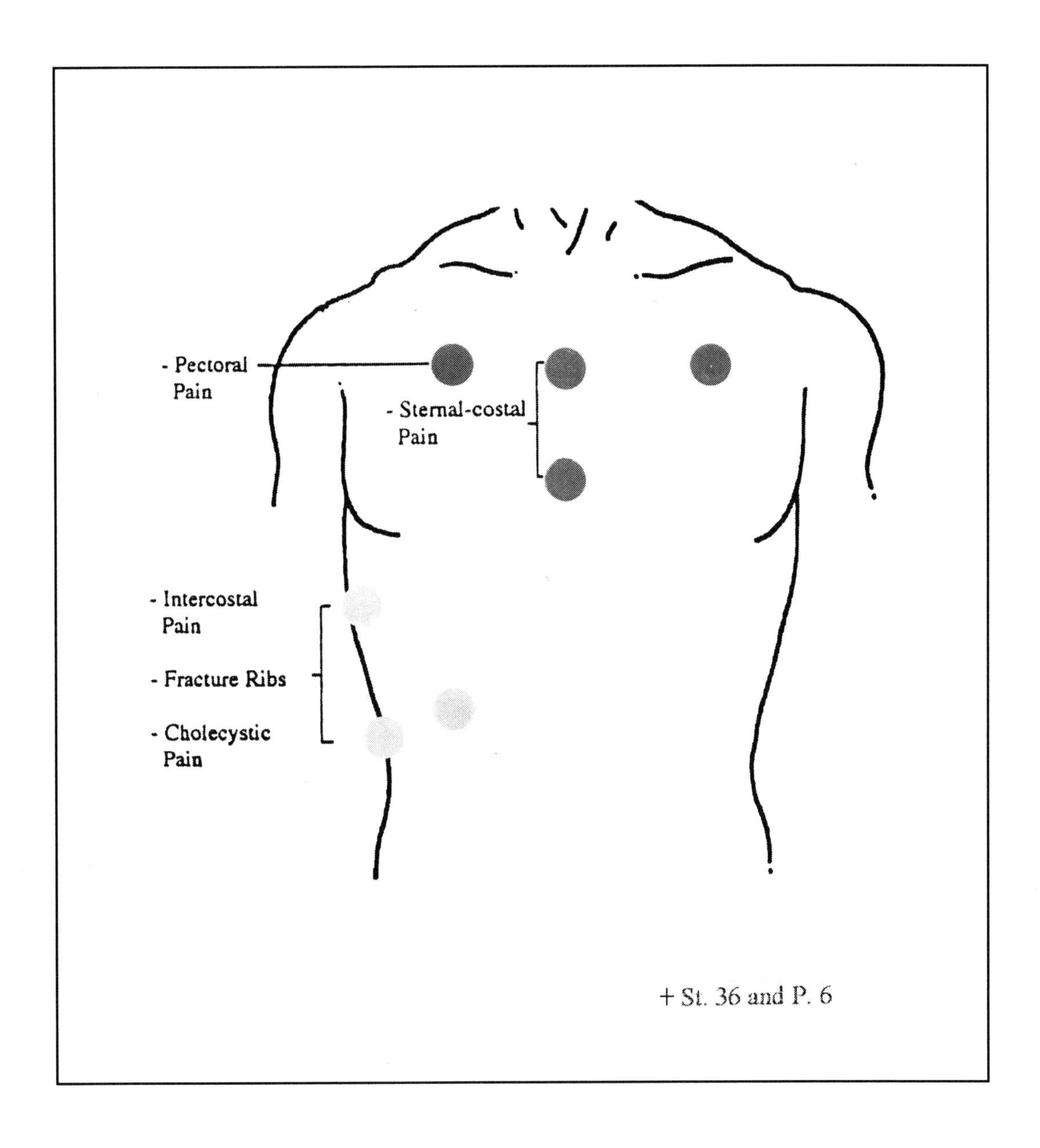

Musculoskeletal Disorders of the Chest and Abdomen
2f) Diagram 23.
Stomach Ache

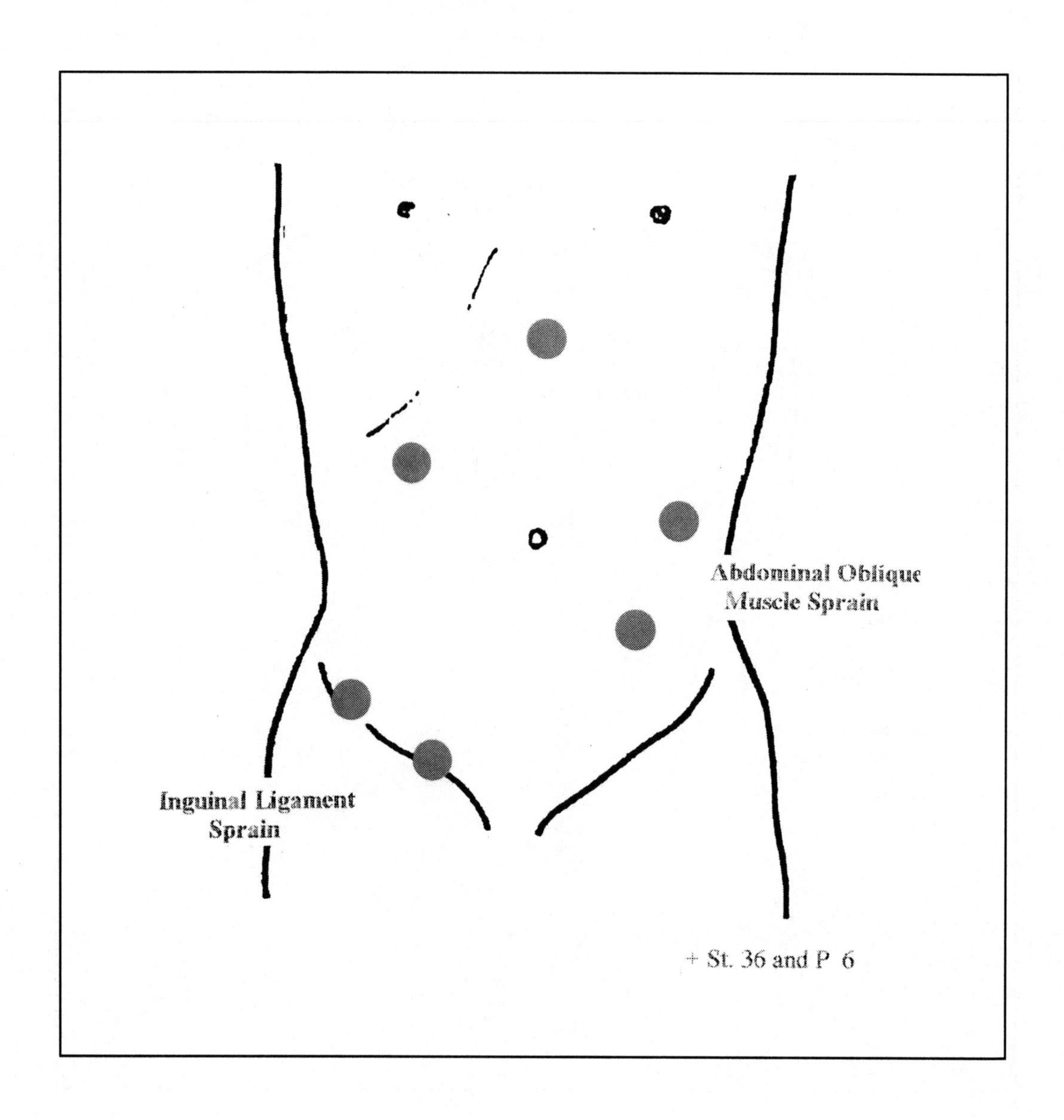

Musculoskeletal Disorders of the Chest and Abdomen
2f) Diagram 24.
Menstrual Pain

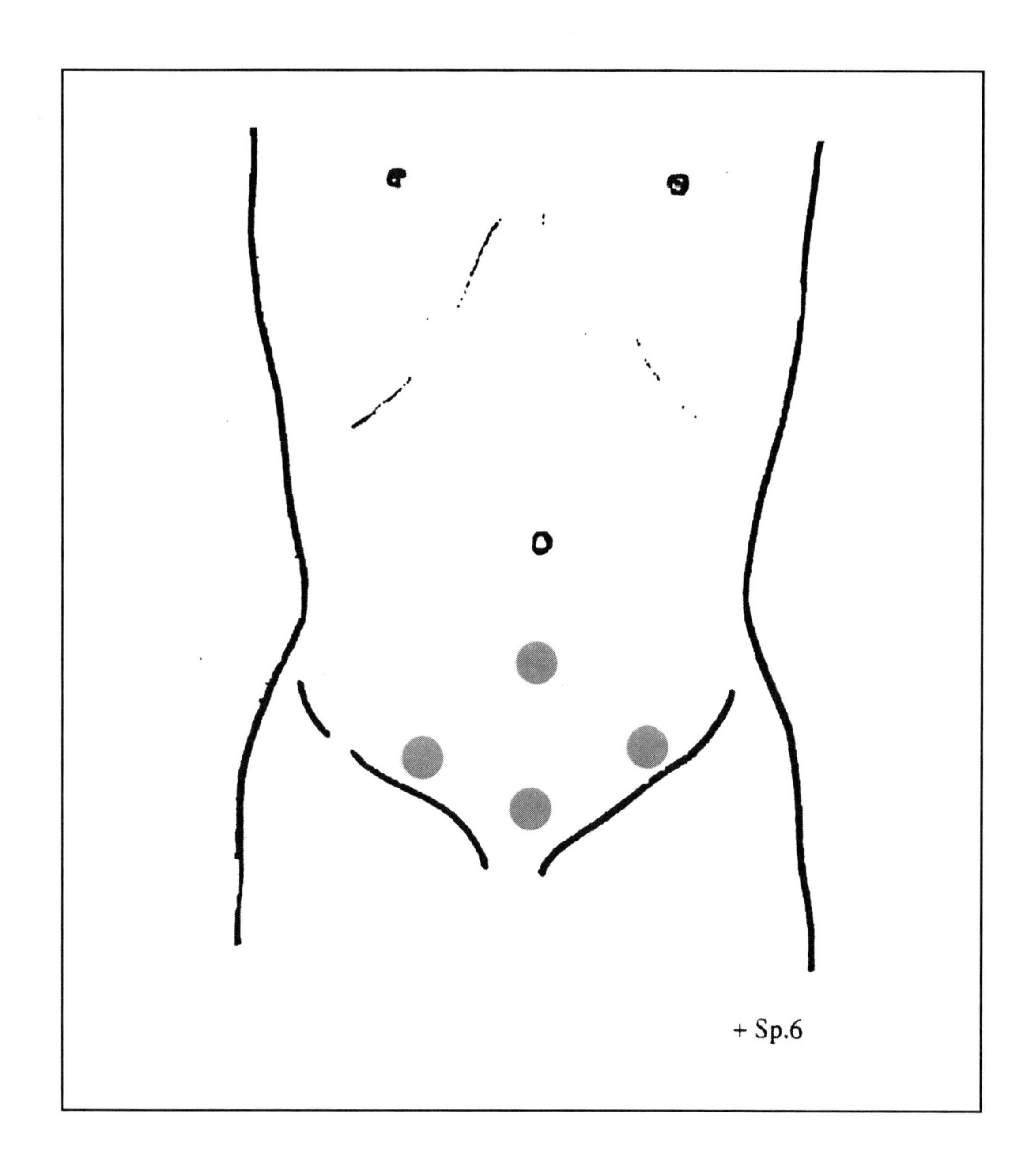

Musculoskeletal Disorders of the Lower Back
2 g) Table 12.
Common Musculoskeletal Disorders of the Lower Back

Lumbo-Sacral Spine	**Mechanical Disorders:** Disc protrusion Facet syndrome Spondylolysis Spondylolisthesis
	Functional Disorders: Non-specific inflammation Specific inflammation
Soft Tissues	**Sprains and Strains:** Ligamentous Muscular Tendinous.

Musculoskeletal Disorders of the Lower Back
2 g) Table 13.
Low Back Pain Lesions: Symptoms, Signs and Acupuncture Points

Level of Lesion	Pain	Numbness	Weakness & Atrophy	Diminished DTR	Acupuncture Point (Fig)
L.4	Low back, hip posterior, thigh, anterior leg	Anterio-medial thigh and knee	Quadriceps	Knee jerk	B1 25 (6A, B) GV 3
L.5	Over sacral-iliac joint, hip, lateral thigh and leg	Lateral leg, web of great toe	Dorsiflexion of great toe and foot, difficulty walking on heels	Posterior tibial reflex	B1 25 (6A) GV3A (6A, B) (midpoint between L.5 and S.1 vertebrae)
S.1	Over sacral-iliac joint, hip, posterior-lateral thigh and leg to heel	Back of calf; lateral heel, foot and toe	Plantar flexion of foot and great toe	Ankle jerk	B1 27 (6A,B) GV 3B (midpoint between S.1 and S.2)

Musculoskeletal Disorders of the Lower Back
2g) Diagram 25.

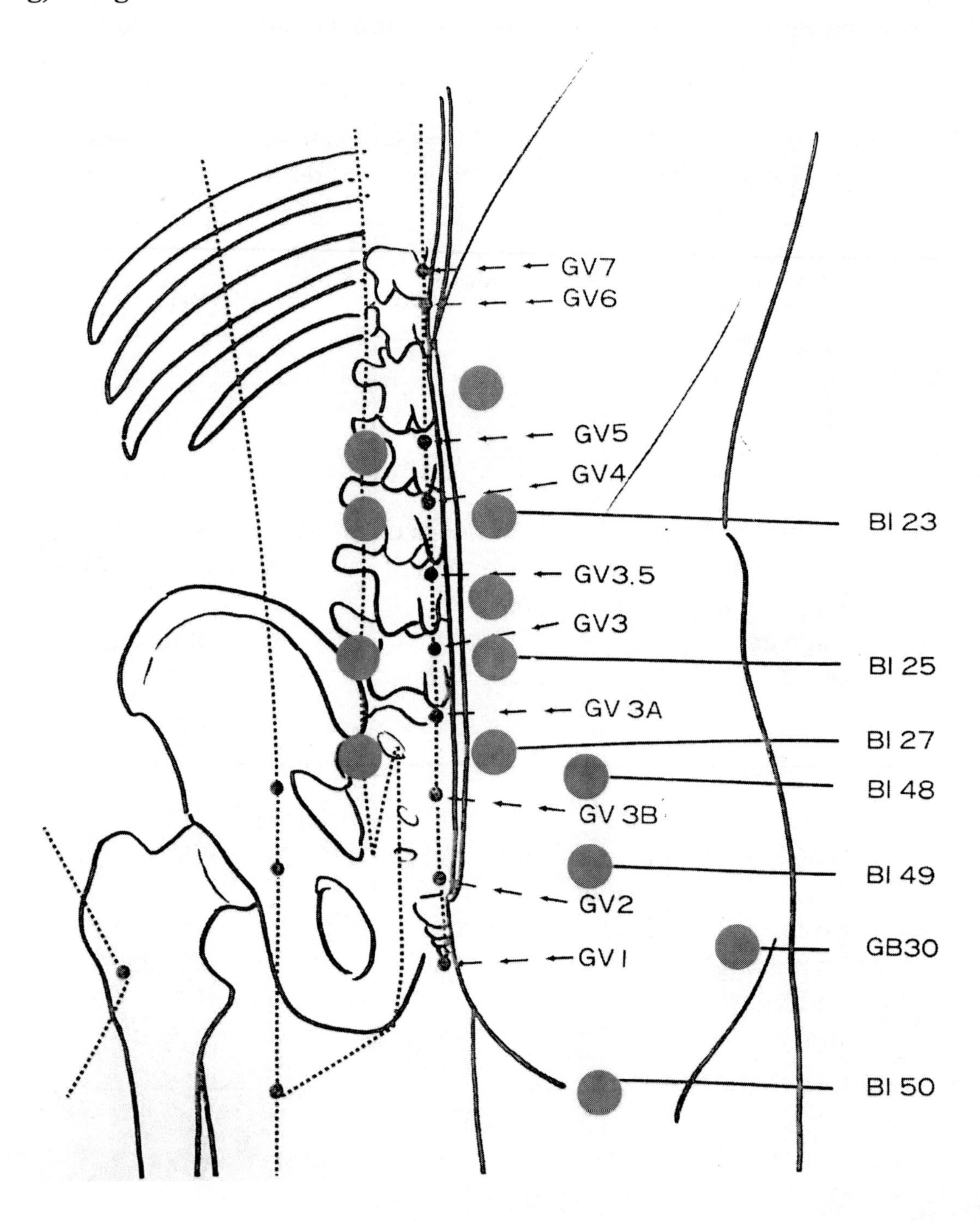

Musculoskeletal Disorders of the Lower Back
2g) Diagram 26.

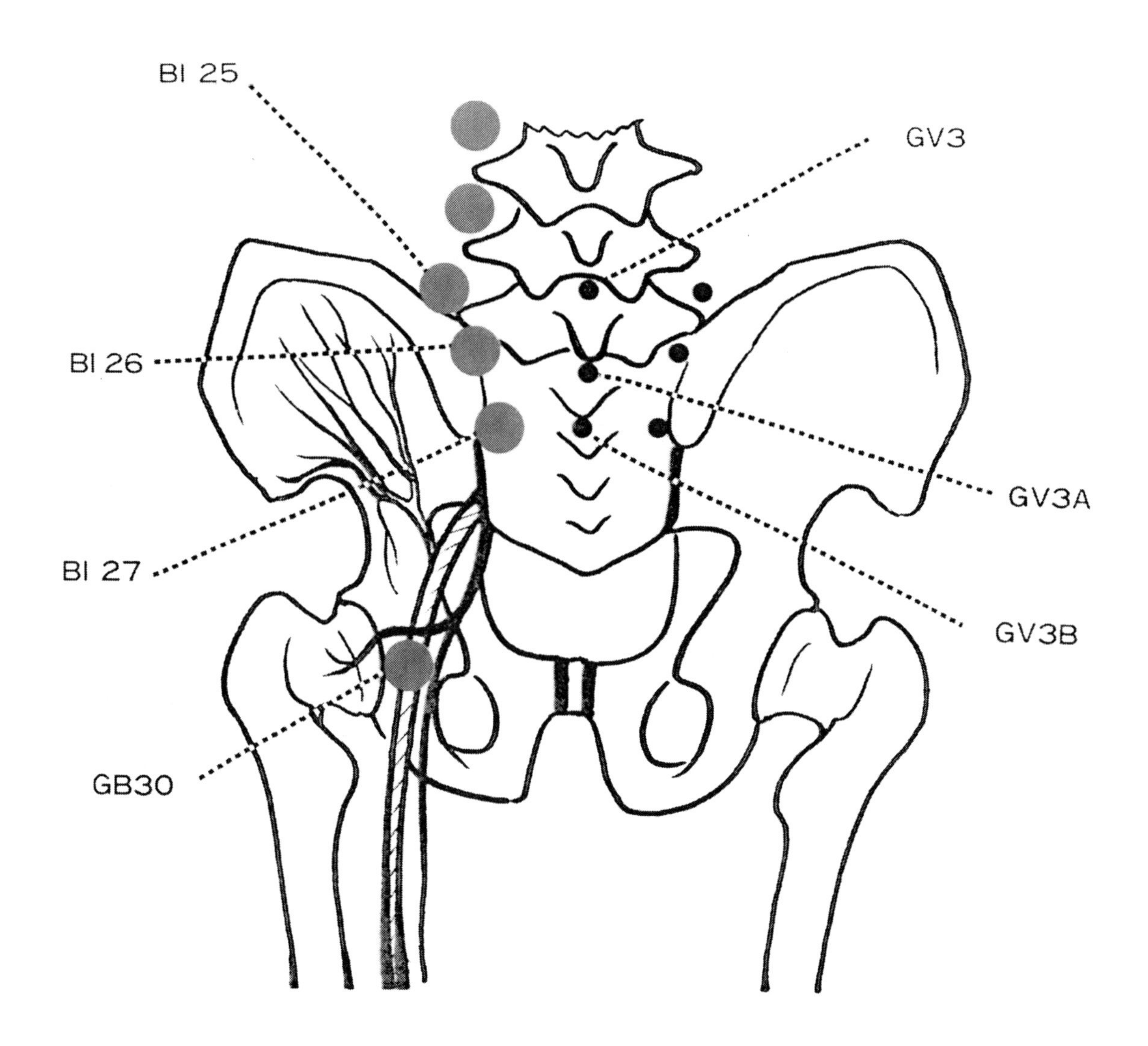

Musculoskeletal Disorders of the Lower Back
2g) Diagram 27.
Low Back Pain: Degenerative Disc, Sciatica, Bursitis

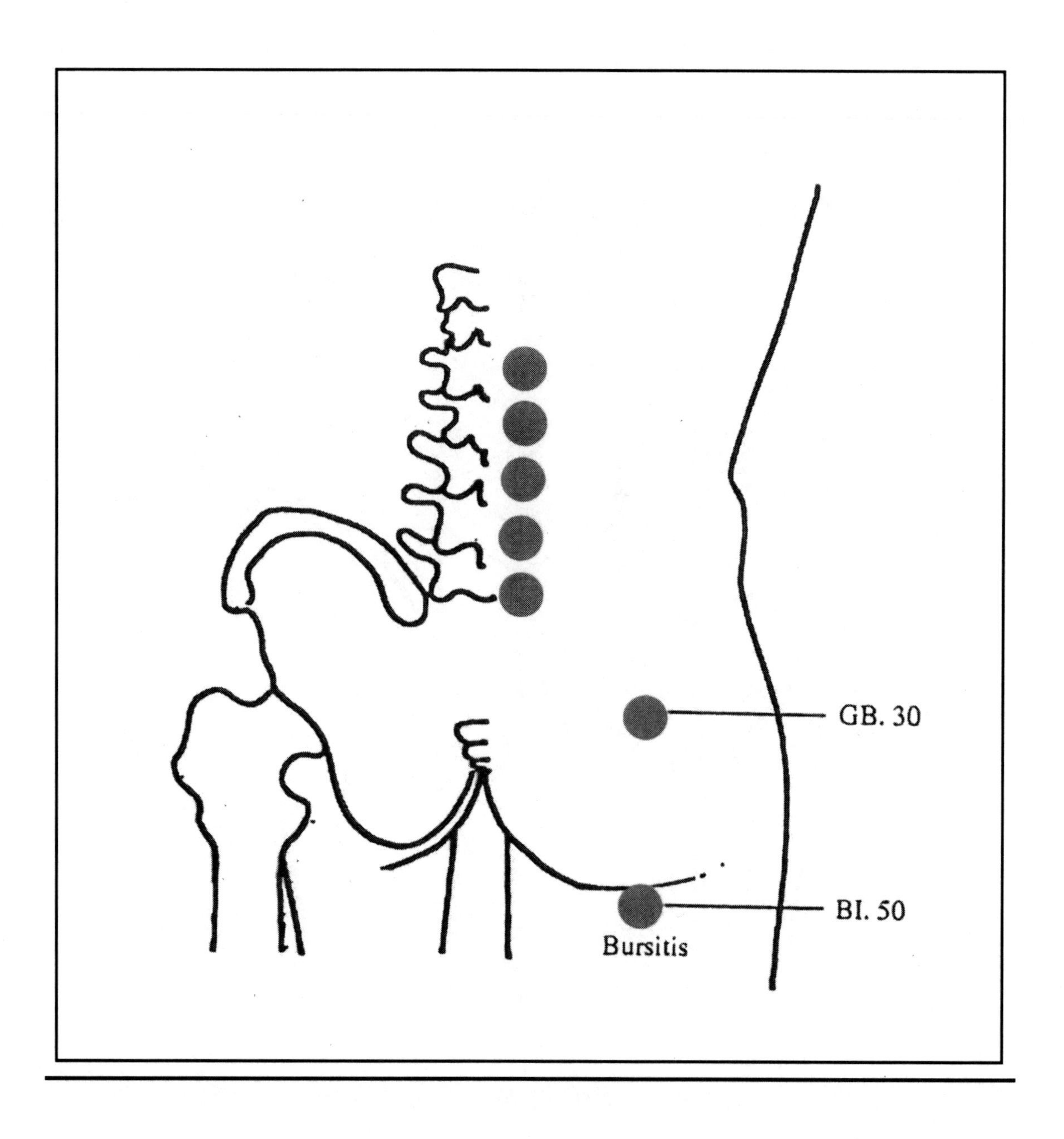

Back Pain

Spinal Muscles

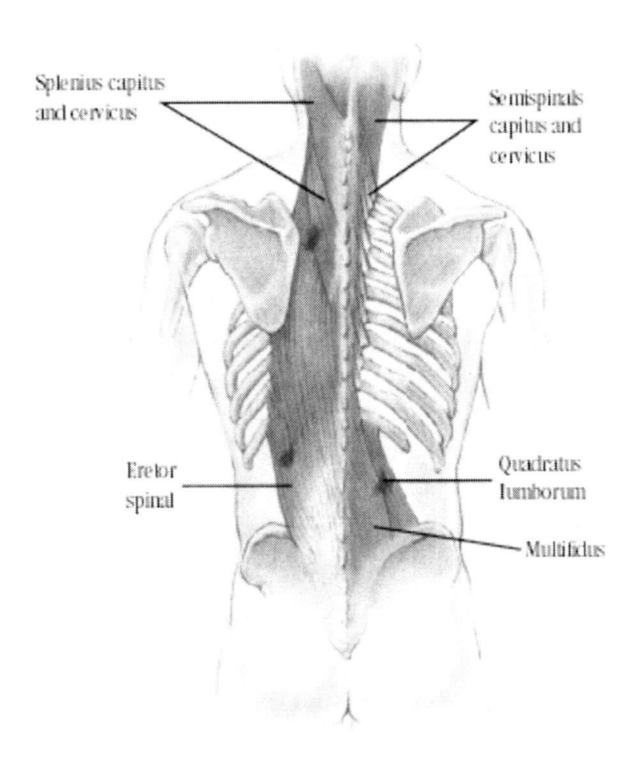

Musculoskeletal Disorders of the Lower Back
2 g) Table 14.
Acupuncture Points for Sciatica

Acupuncture Point (Fig)	Location	Anatomy
GB 30	Outer 1/3 on a line joining the greater trochanter of the femur and the hiatus of the sacrum	Sciatic nerve L.4, 5, S.1, 2, 3
B1 54	Midpoint of the popliteal fossa	Tibial nerve L.4, 5, S.1, 2, 3

Musculoskeletal Disorders of the Lower Back
2 g) Table 15.
Acupuncture Points for
Referring Pain and Paresthesia in the Leg or Foot

Acupuncture Point (Fig)	Location	Anatomy
K3 Medial	Midpoint between the tip of medial malleolus and the tendon calcaneus	Saphenous nerve L.4 Tibial nerve L.4, 5, S.1, 2, 3
B1 60 Lateral	Midpoint between the tip of lateral malleolus and the tendon calcaneus	Lateral sural nerve L.5 Peroneal nerve L.4, 5, S.1

Musculoskeletal Disorders of the Lower Back
2g) Diagram 28.
Sciatica

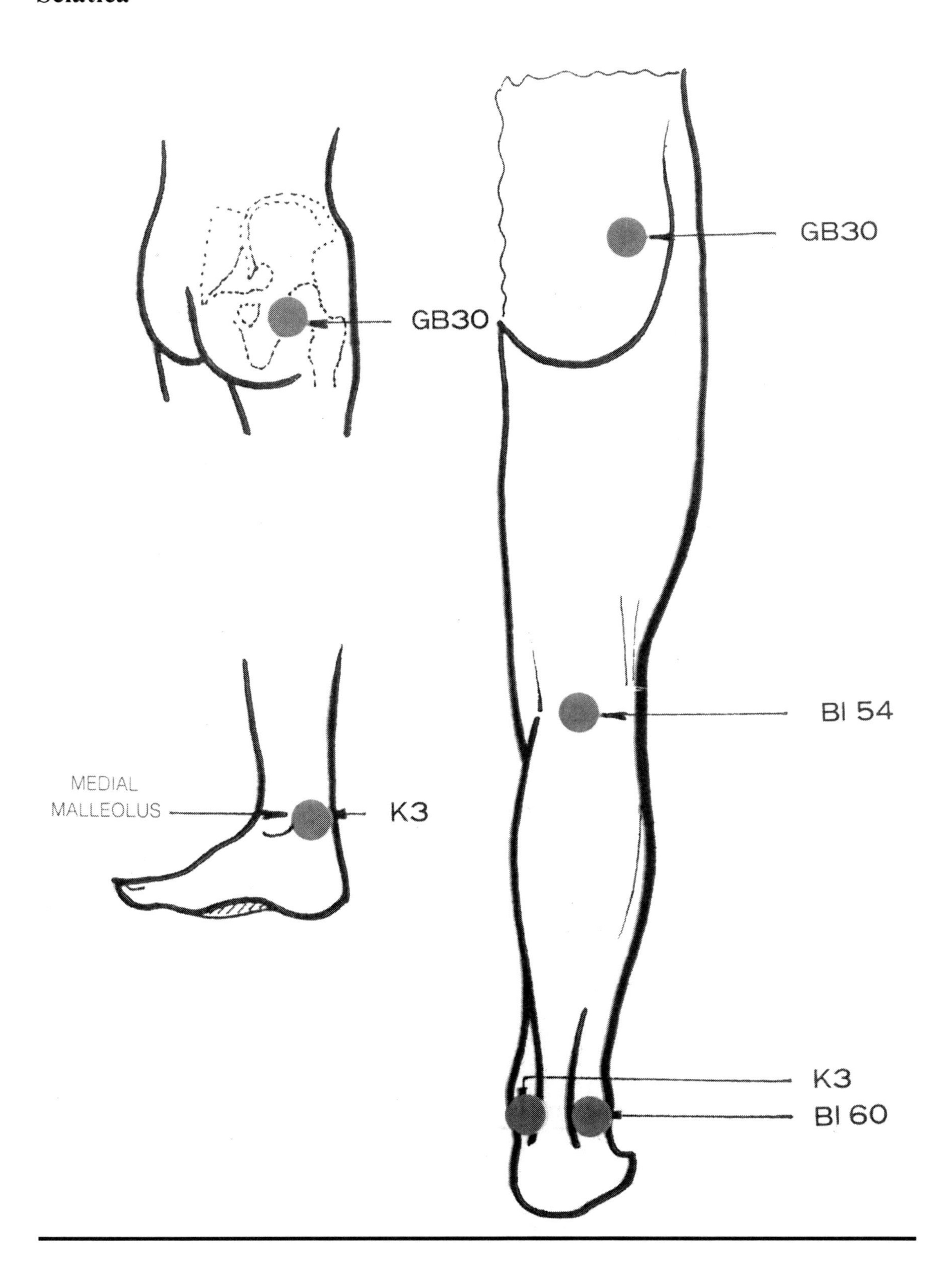

Musculoskeletal Disorders of the Lower Back
2g) Diagram 29.
Sciatica, Posterior Knee and Ankle Pain

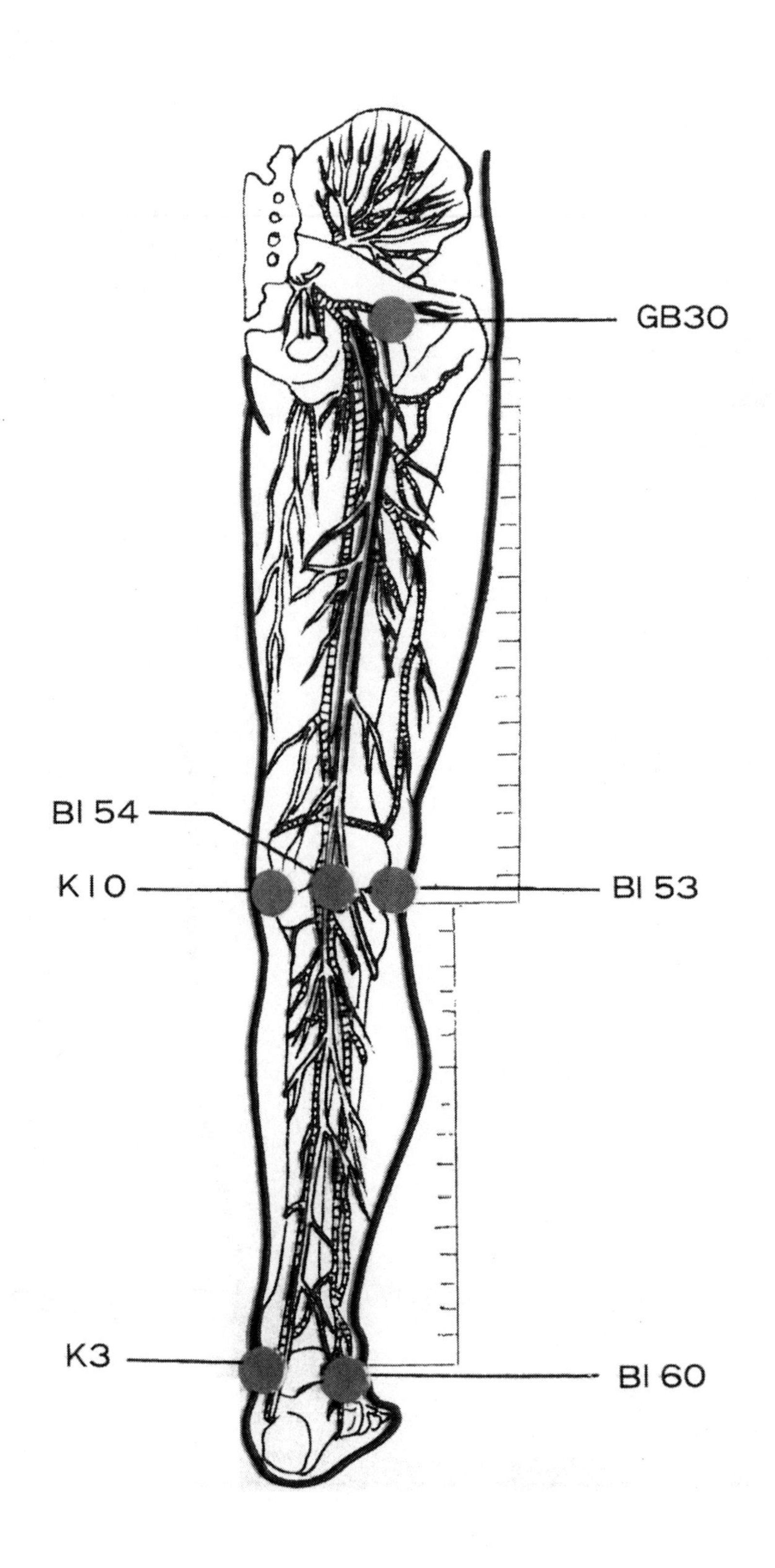

Musculoskeletal Disorders of the Lower Back
2g) Diagram 30.
Low Back Pain: Osteoarthritis, Sprain and Strain
Osteoporosis, Degenerative Disc, Muscular Sprain

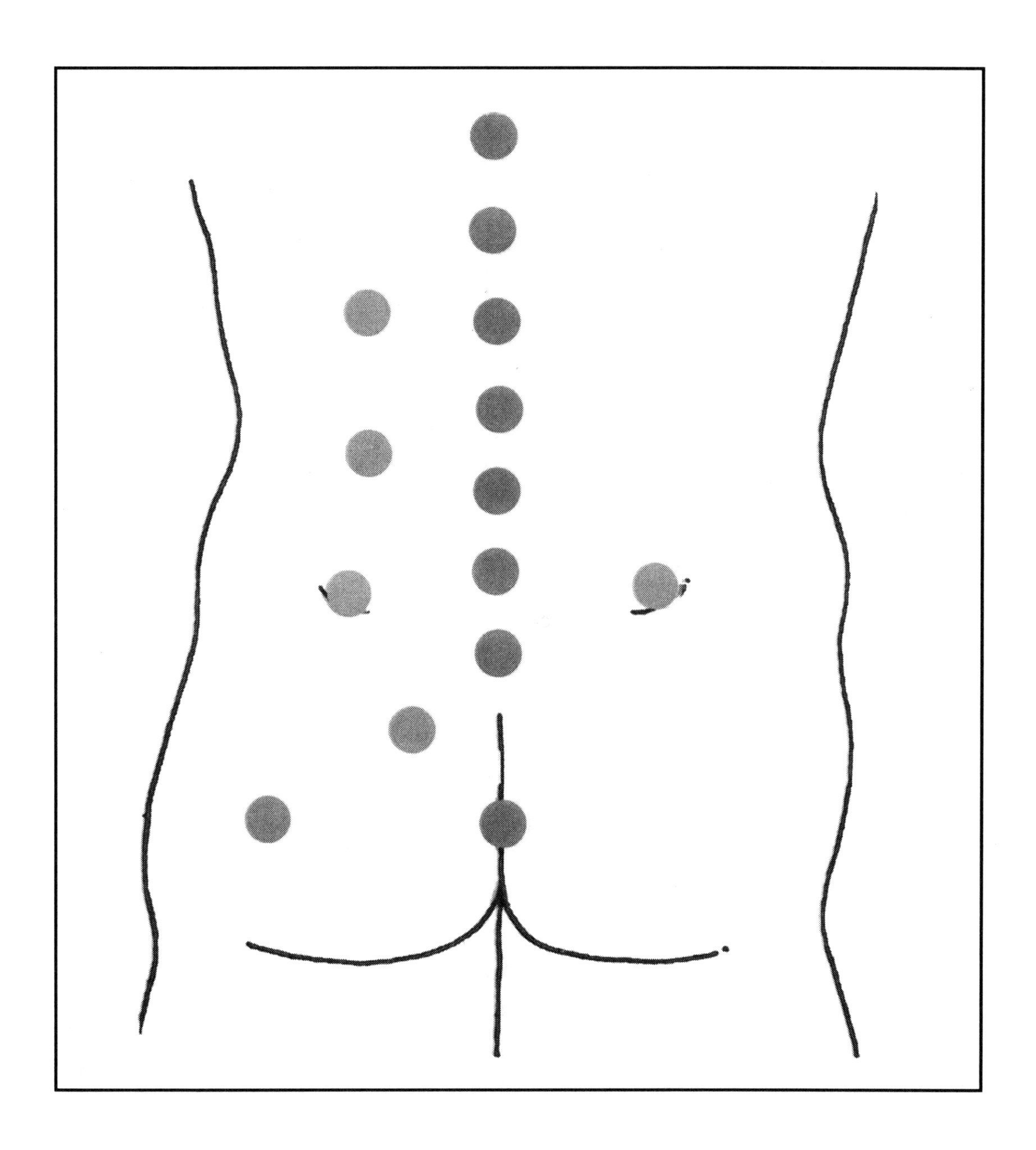

Musculoskeletal Disorders of the Hip
2 h) Table 16.
Common Musculoskeletal Disorders of the Hip

Non-specific Inflammation	• Osteoarthritis • Traumatic synovitis • Gluteal bursitis • Adductor sprain (Rider's sprain) • Psoas strain • Ilio-tibial fascia strain
Specific Inflammation	• Inflammation • Rheumatoid arthritis • Ankylosing spondylitis

Musculoskeletal Disorders of Hip
2 h) Table 17.
Acupuncture Points for Disorders of the Hip

	Acupuncture Point (Fig)	**Location**	**Anatomy**
Anterior disorders: • Osteoarthritis • Psoas strain	Sp 12	Proximal to midpoint of inguinal ligament	Femoral nerve Femoral artery
Posterior disorders: • Osteoarthitis • Bluteal bursitis	B1 49	3” lateral to 3rd sacral vertebra	Inferior guteal nerve
Lateral disorders: • Osteoarthritis • Ilio-tibial fascia strain	B1 48 GB 31	3” lateral to 2nd sacral vertebra Lateral aspect of thigh, 7” above the knee	Superior gluteal nerve Ilio-tibial band femoral nerve
Medial disorders: • Adductor sprain	Liv 11	Below pubic tubercle insertion of adductor longus	Medical cutaneous nerve Obturator nerve

Musculoskeletal Disorders of the Hip
2h) Diagram 31.
Hip Pain – a) Posterior and Lateral

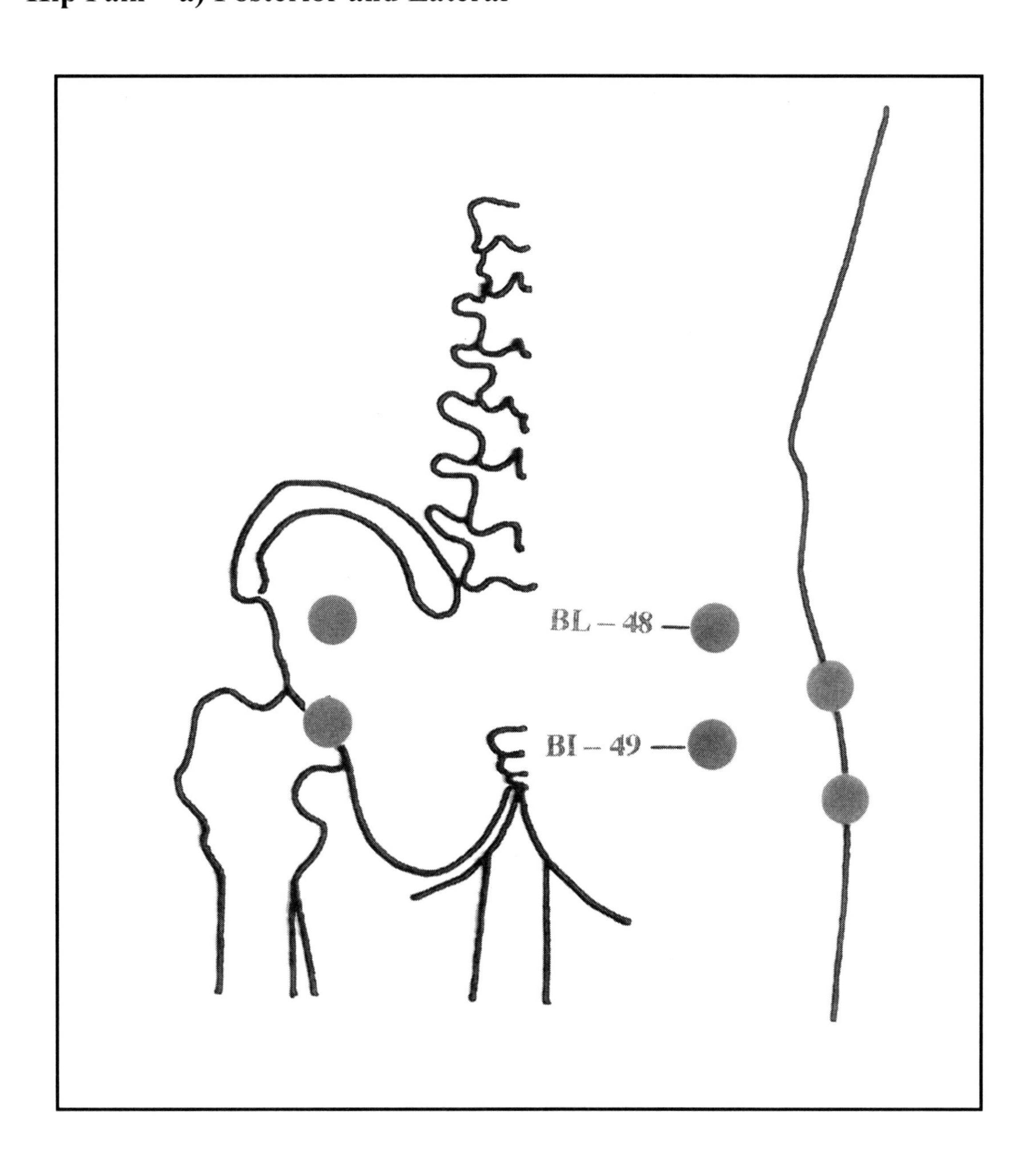

Musculoskeletal Disorders of the Hip
2h) Diagram 32.
Hip and Groin Pain – a) Lateral Anterior Hip Pain, Inguinal, Obturator, Adductor Muscular Spain

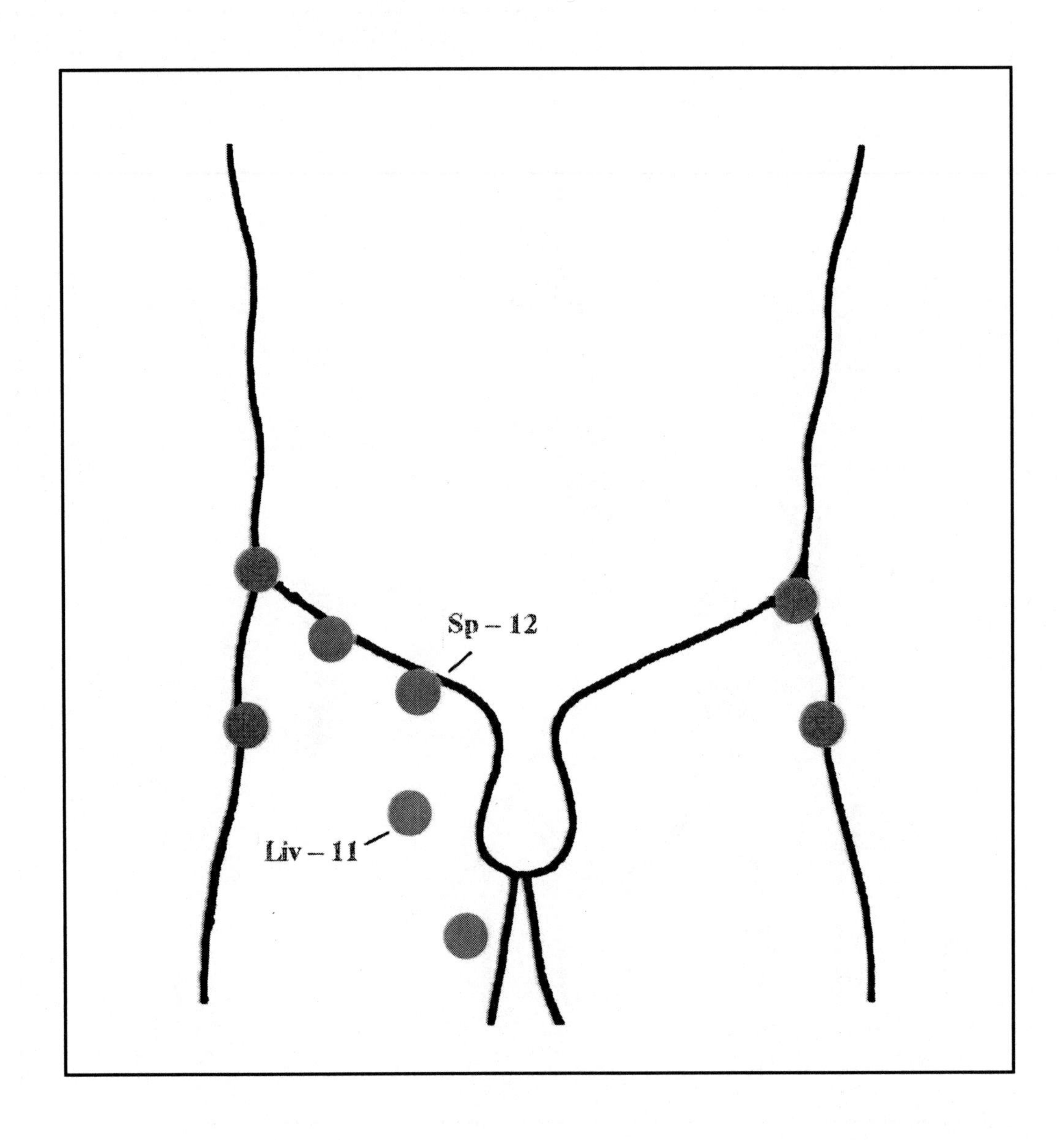

Musculoskeletal Disorders of the Hip
2h) Diagram 33.
Quadriceps Pain - Sprain and Strain

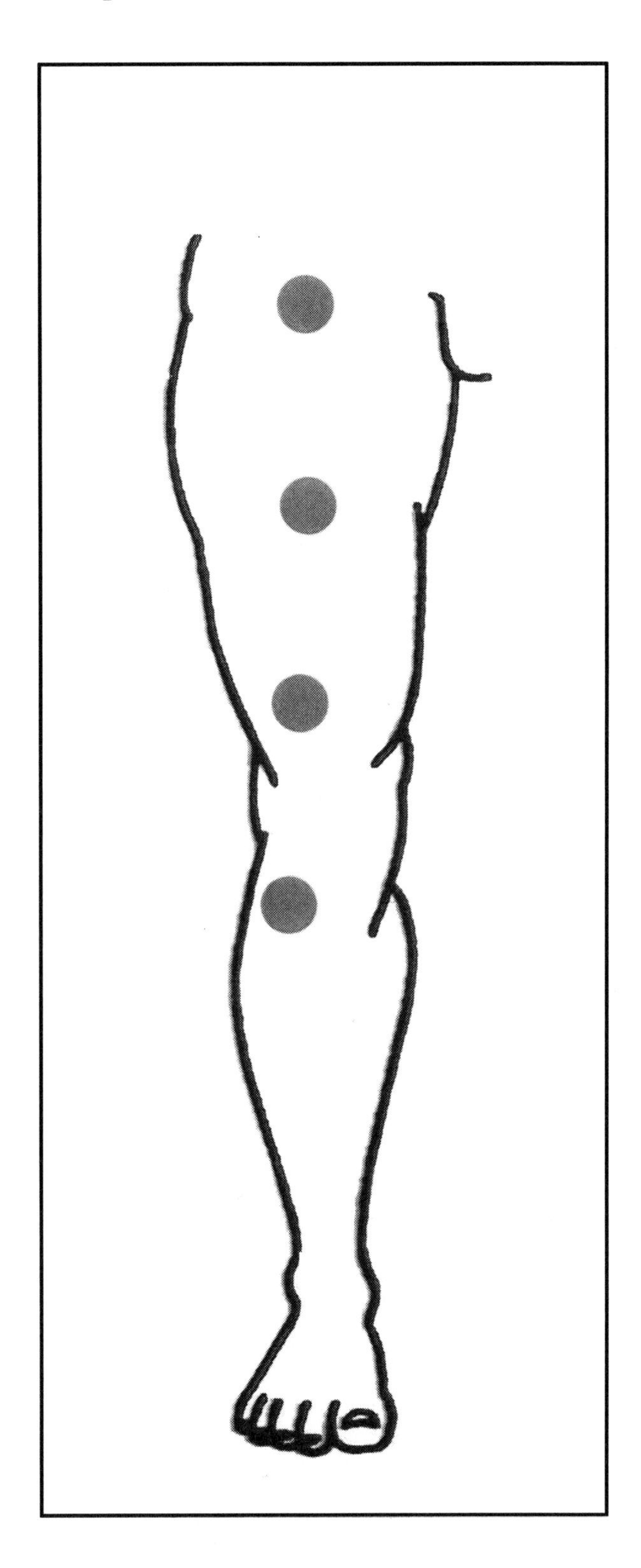

Musculoskeletal Disorders of the Hip
2h) Diagram 34.
Hamstring Pain - Sprain and Strain

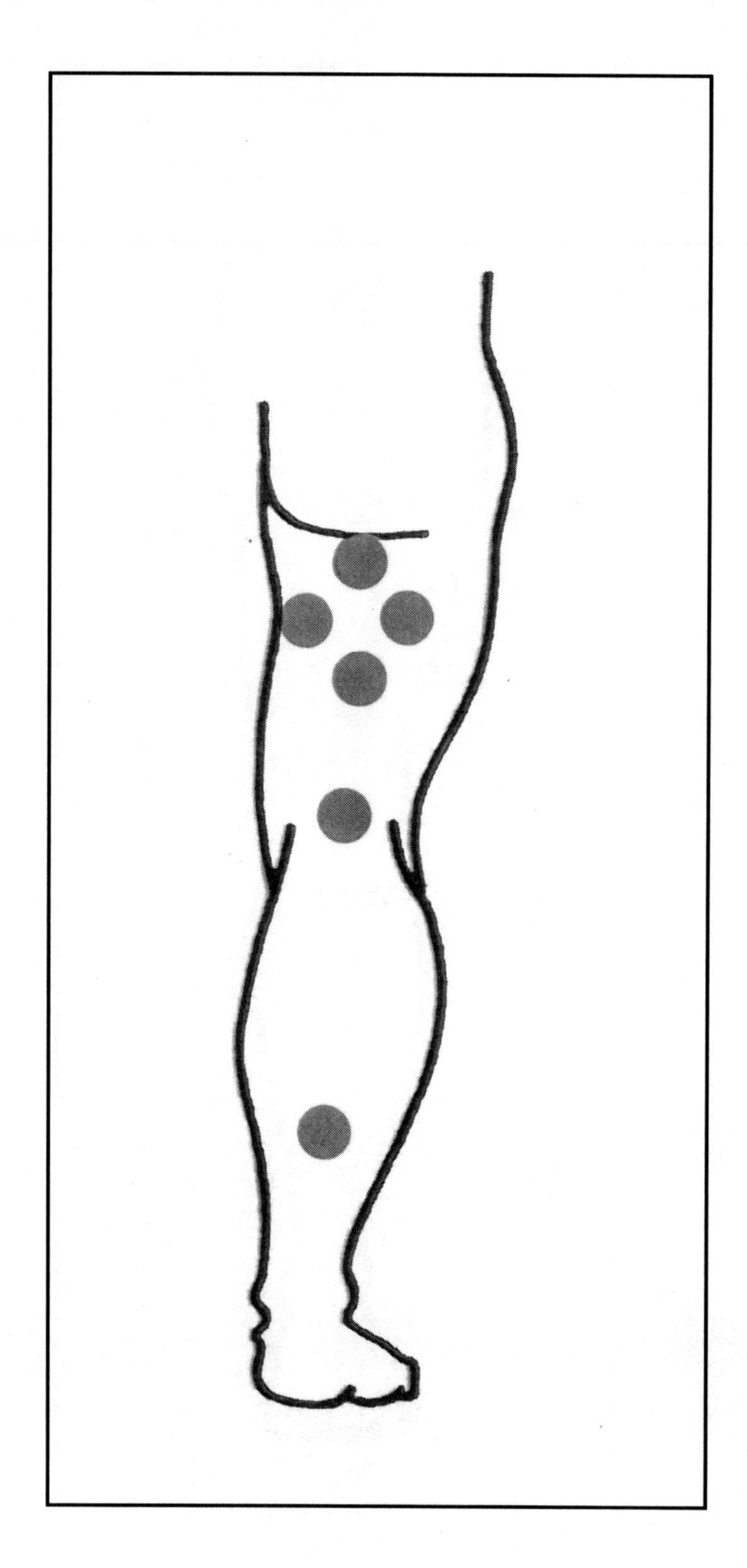

Musculoskeletal Disorders of the Hip
2h) Diagram 35.
Leg Pain

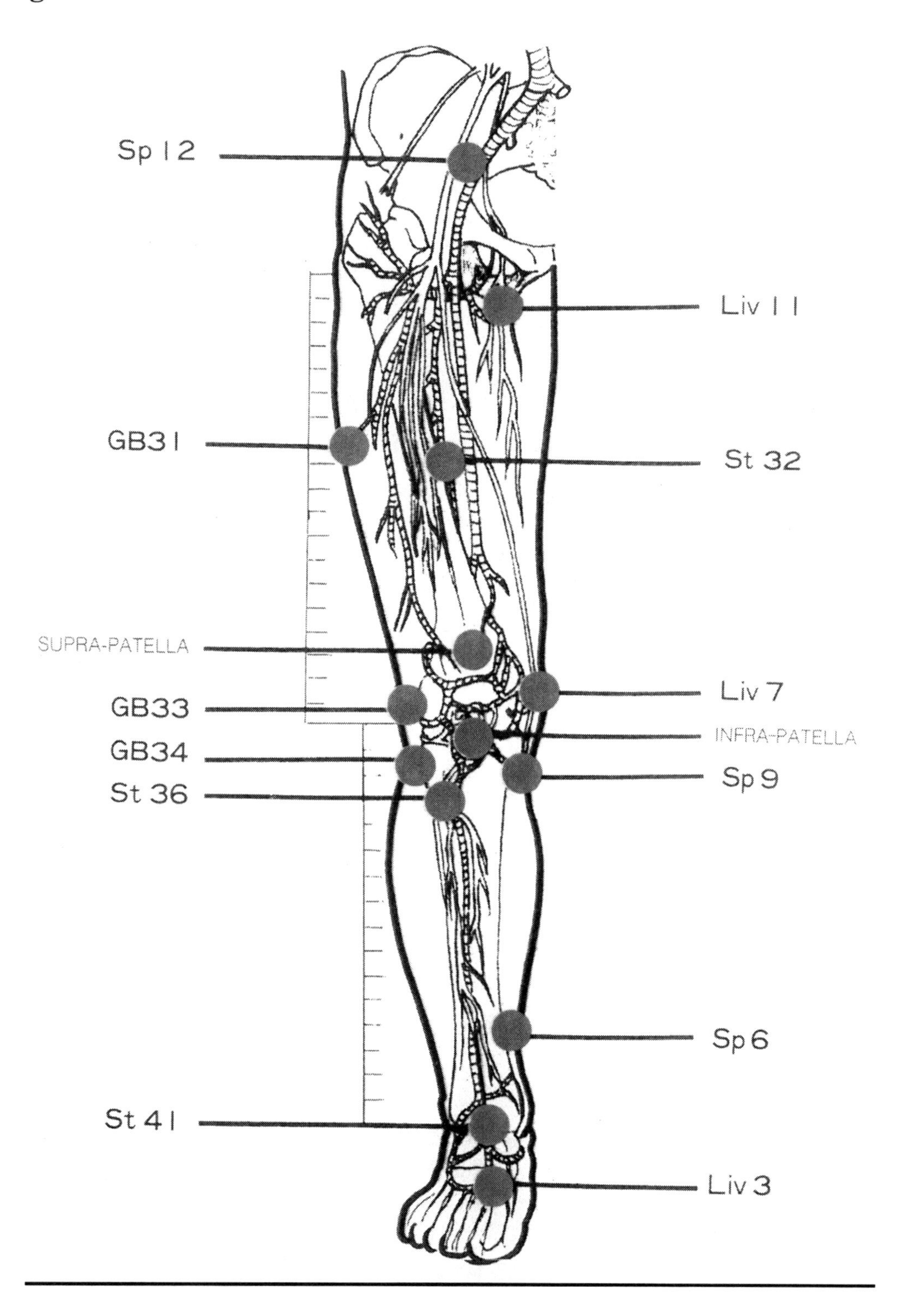

Muscle Tear

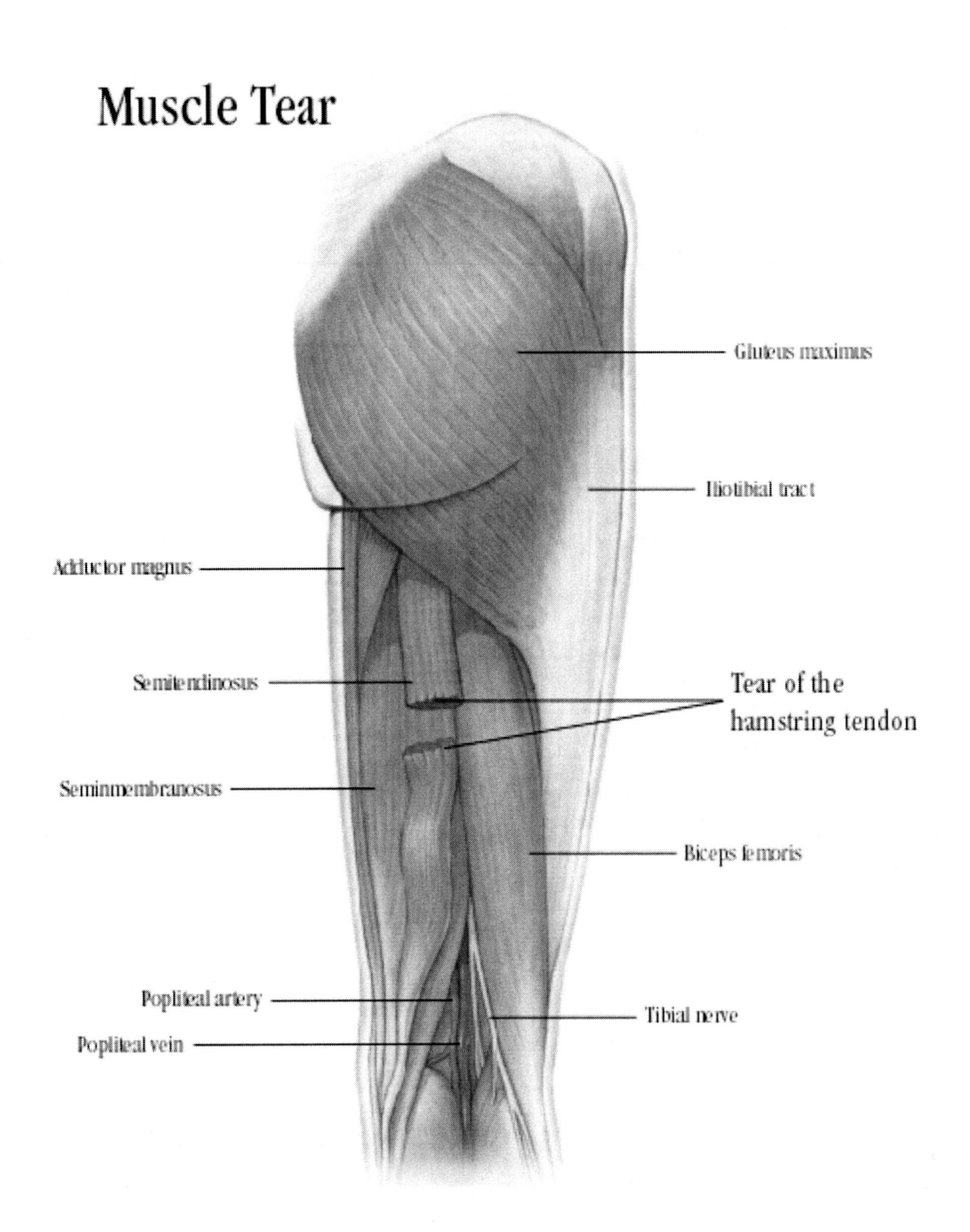
Gluteus maximus
Iliotibial tract
Adductor magnus
Semitendinosus
Tear of the
hamstring tendon
Seminmembranosus
Biceps femoris
Popliteal artery
Tibial nerve
Popliteal vein

Musculoskeletal Disorders of the Knee
2 i) Table 18.
Common Musculoskeletal Disorders of the Knee

<table>
<tr><td>Non-specific Inflammation</td><td>• Osteoarthritis
• Traumatic synovitis
• Collateral ligamentous strain
- Medial
- Lateral
• Cruciate ligamentous strain
- Anterior
- Posterior
• Meniscus impairment
- Medial
- Lateral
• Patellar chondritis
• Tendinitis
- Quadriceps
- Biceps femoris
- Hamstrings</td></tr>
<tr><td>Specific Inflammation</td><td>• Rheumatoid arthritis
• Infectious arthritis
- e.g., Reiter's syndrome</td></tr>
</table>

Musculoskeletal Disorders of the Knee
2 i) Table 19.
Acupuncture Points for Knee Disorders

Disorder	Acupuncture Point (Fig)	Location	Anatomy
Arthritis - Osteoarthritis - Rheumatoid arthritis - Traumatic synovitis	Sp 9 GB 34	- Below the medial tibial condyle -Anterior to the fibular neck	- Tibial nerve Peroneal nerve - Recurrent articular nerve
Ligamentous - Medial collateral - Lateral collateral - Anterior cruciate - Posterior cruciate	Liv 7 GB 33 St 35 Knee eye	- Medial knee mid-joint point - Lateral knee mid-joint point - Lateral to infrapatella tendon - Medial to infrapatella tendon	Intra-articular
•Miniscal •- Medial •- Lateral	Knee eye St 35	- Medial to infrapatella tendon - Lateral to infrapatella tendon	Intra-articular
Patella - Chondritis	- Supra patella - Infra patella	- Mid-point above patella - Mid-point below patella	Superpatella tendon Infrapatella tendon
Tendinitis - Quadriceps - Biceps femoris - Hamstrings	St 32 B1 53 K 10	- 6” above patella - Medial to biceps tendon - Popliteal crease between semi-tendinous & semi-membranous	- Quadriceps Musculotendinous junction - Biceps tendon - Hamstrings tendon

ACL Injuries
(Anterior Cruciate Ligament)

Shin Splints

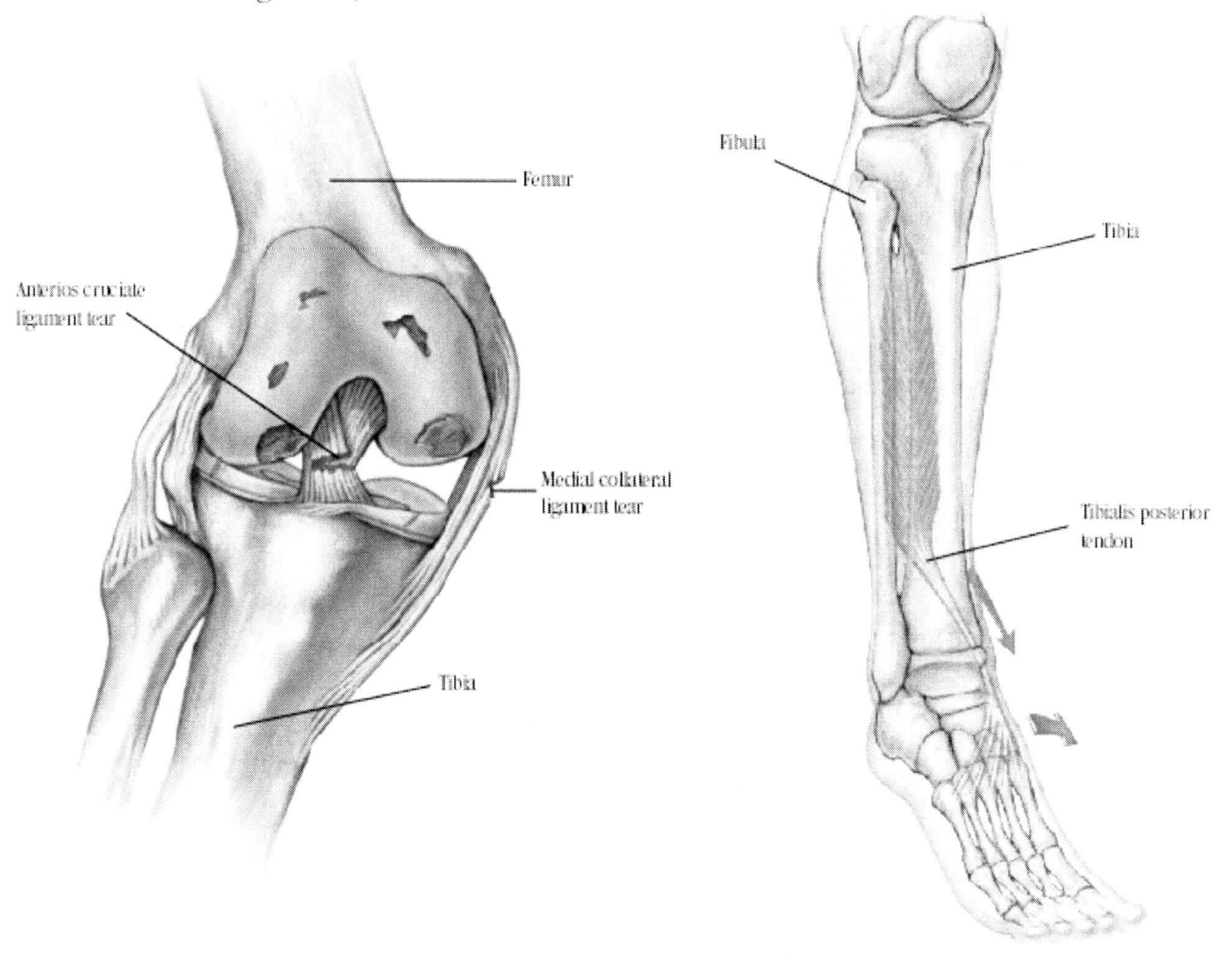

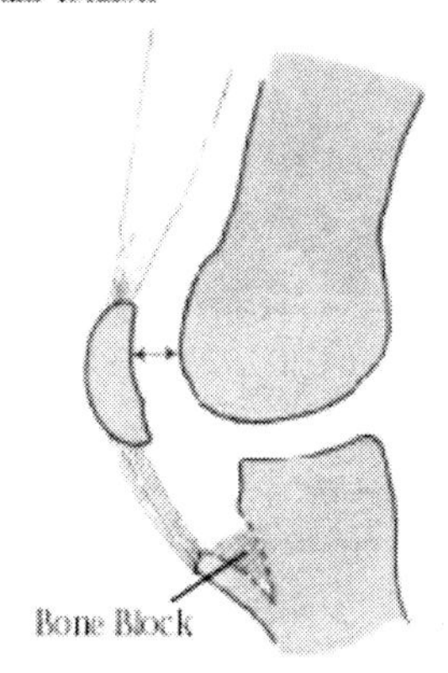

Musculoskeletal Disorders of the Knee
2i) Diagram 36.
Knee Pain

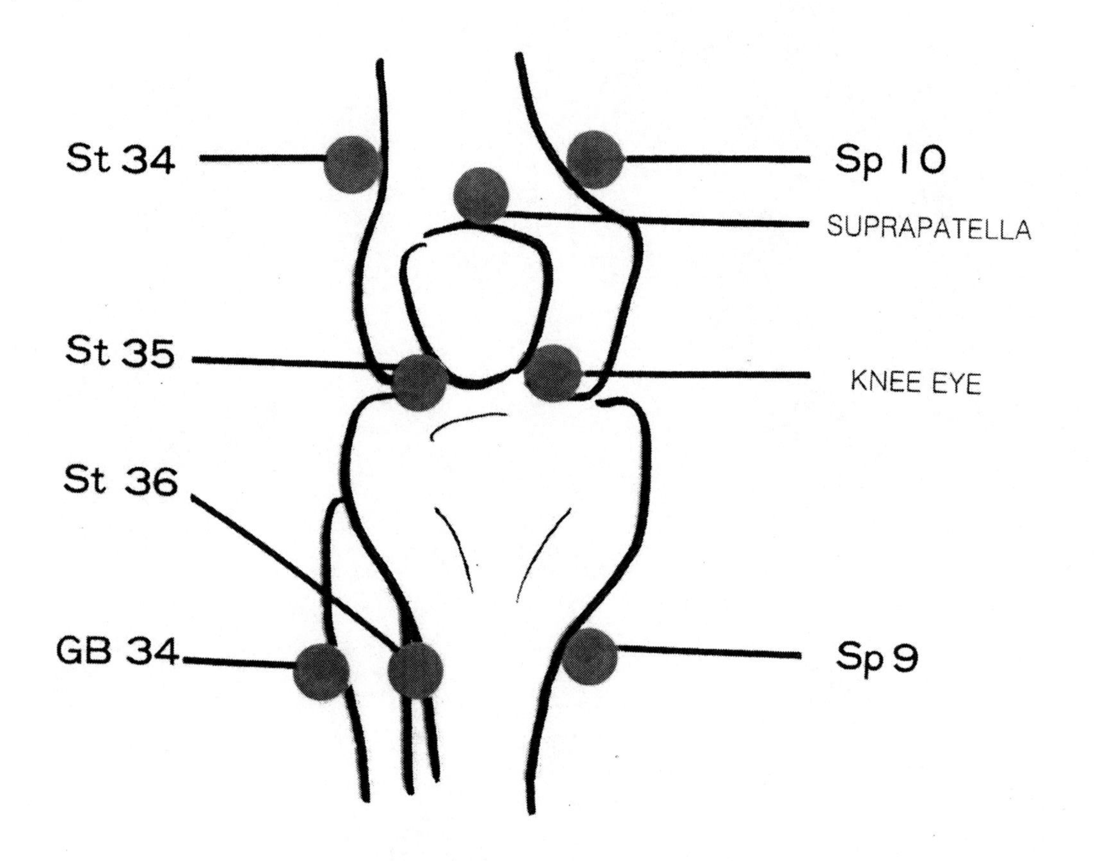

Musculoskeletal Disorders of the Knee
2i) Diagram 37.
Knee Pain – Osteoarthritis, Rheumatoid Arthritis, Chondromalacia, Ligament Sprain

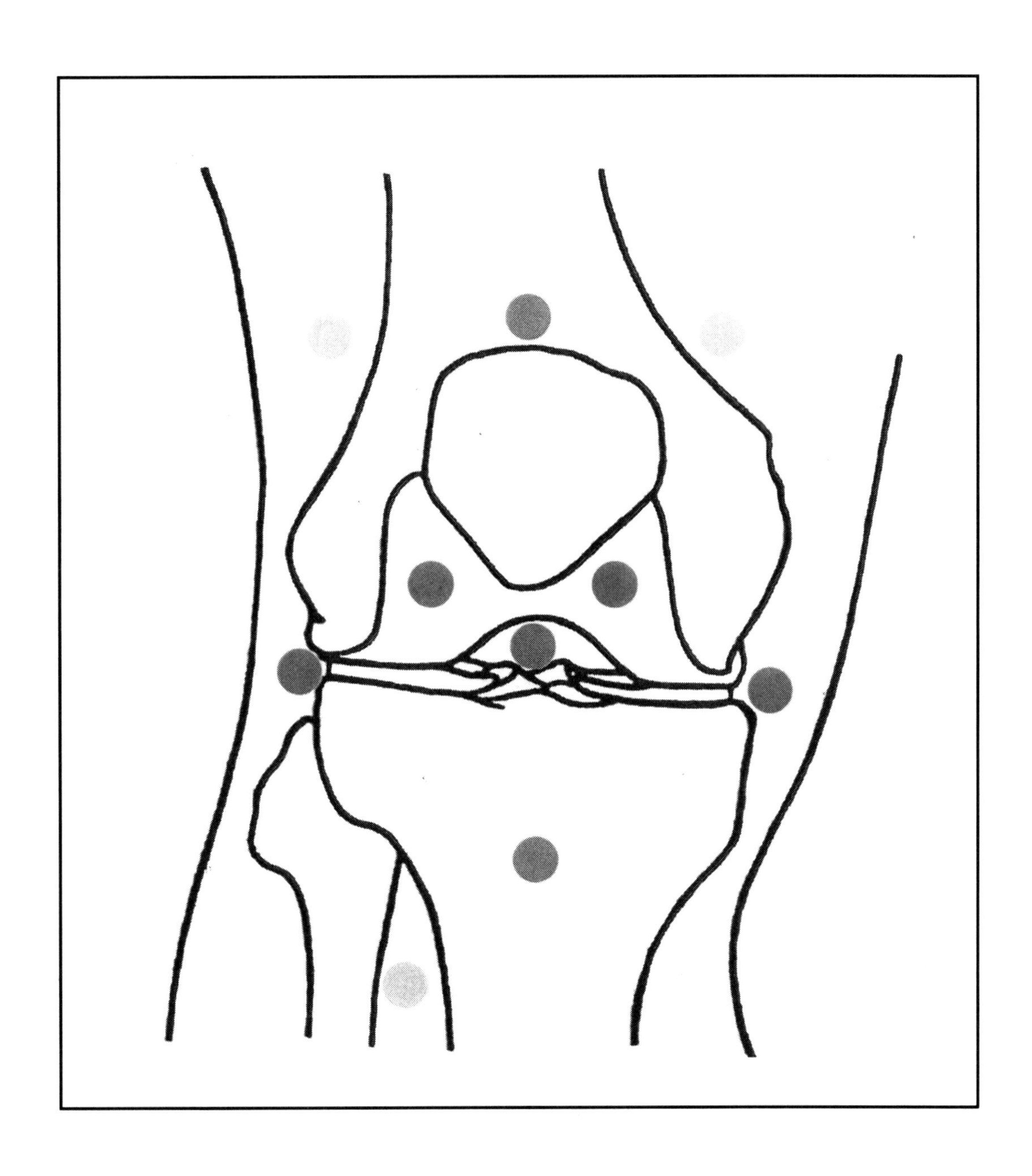

Musculoskeletal Disorders of the Knee
2i) Diagram 38.
Shin Splint - Strain and Sprain, Medial and Lateral

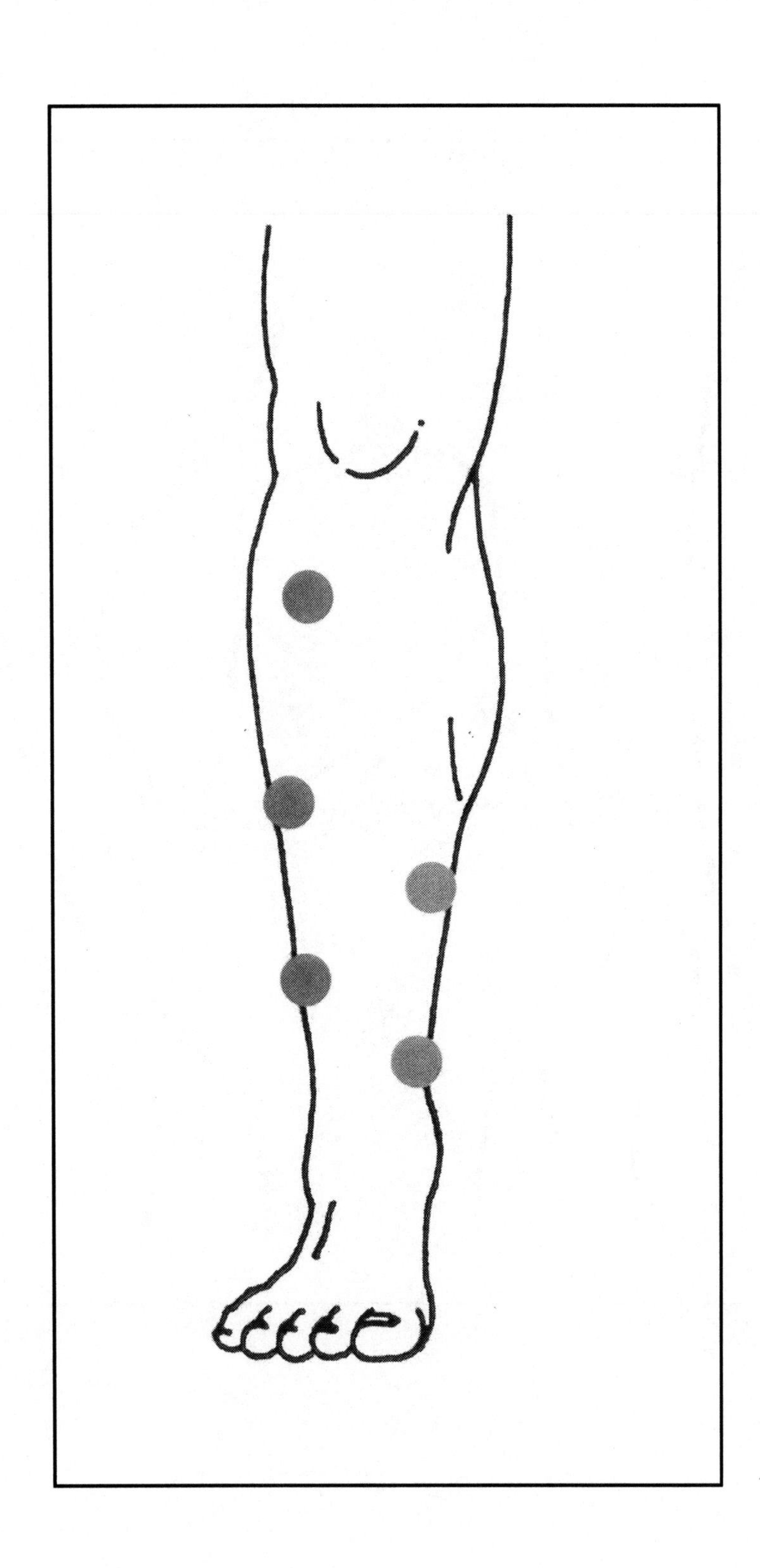

Musculoskeletal Disorders of the Knee
2i) Diagram 39.
Leg Pain

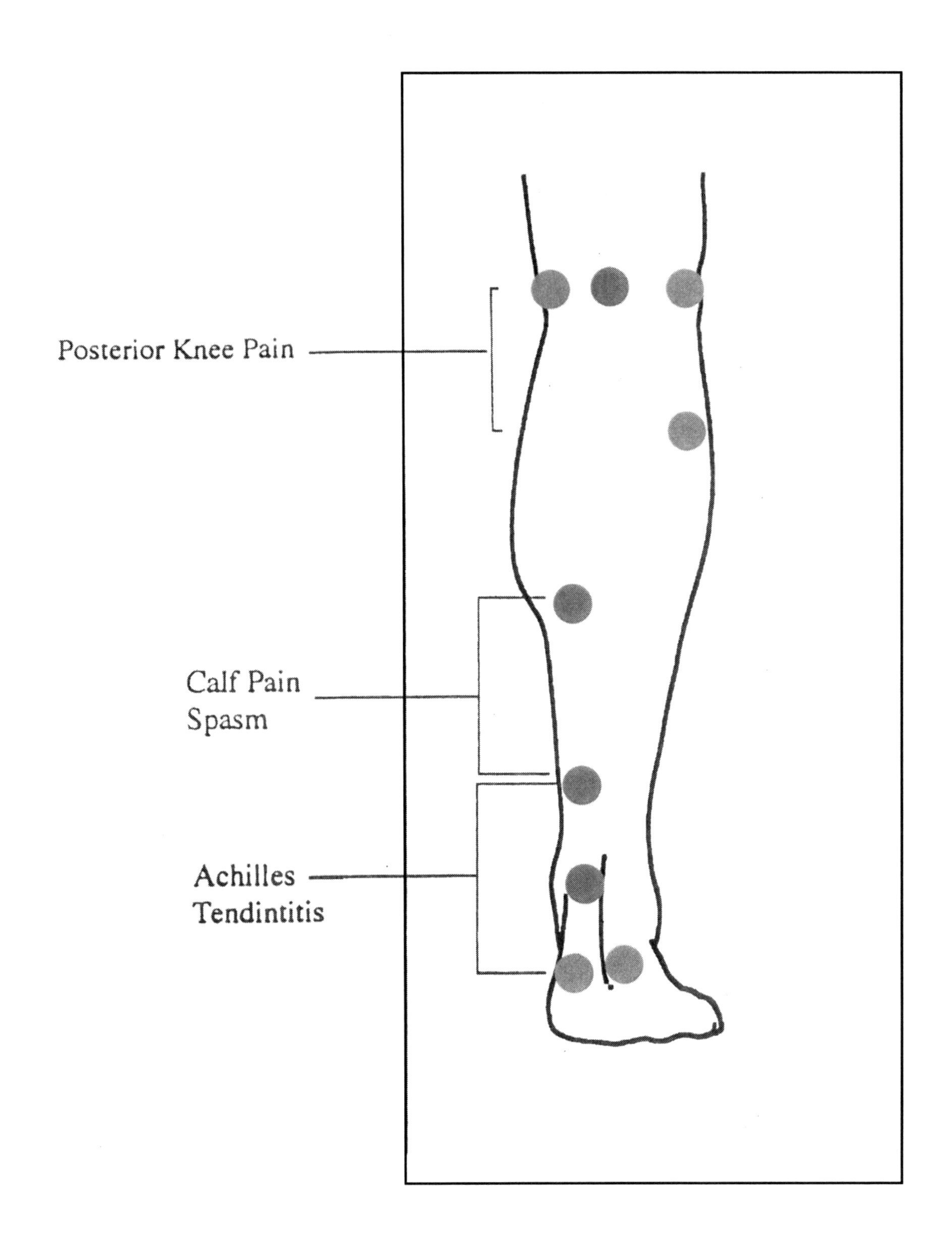

Musculoskeletal Disorders of the Ankles, Feet and Toes
2 j) Table 20.
Acupuncture Points for Disorders of the Ankle

Disorder	Acupuncture Point (Fig)	Location	Anatomy
Medial collateral ligamentous strain	K3	Posterior to medial Malleolus	Saphenous and tibial nerves
Lateral collateral ligamentous strain	B1 60	Posterior to lateral malleolus	Sural nerve
Osteoarthrtis	St 41	In the anterior aspect of the ankle joint, between the tendons of extensor digitorum longus and extensor hallucis longus	Peroneal nerve

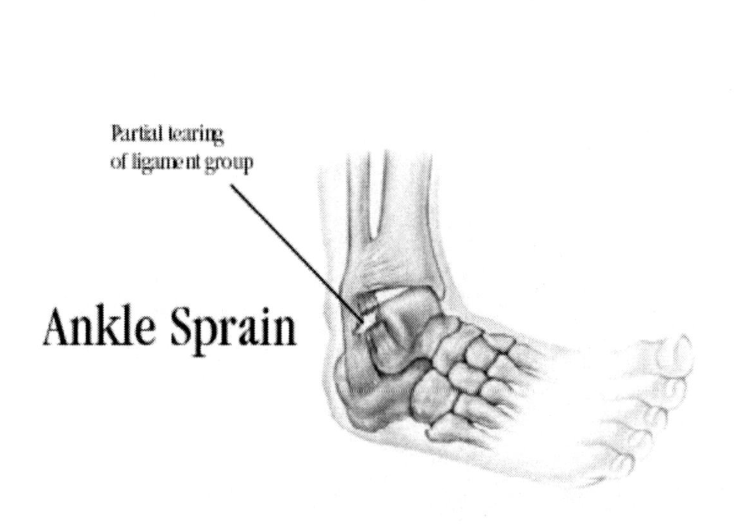

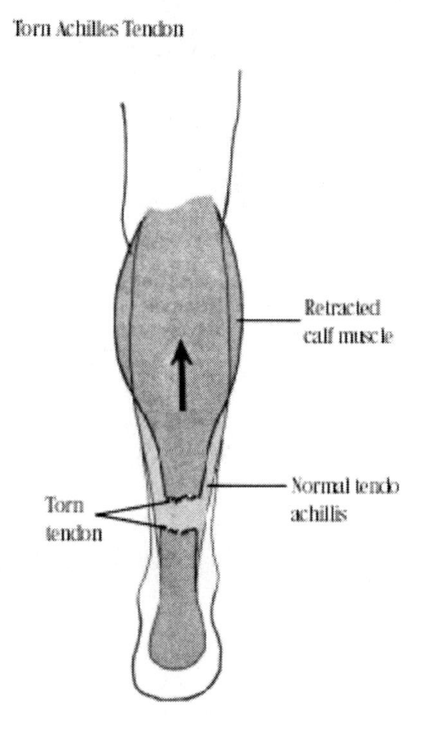

Plantar Fasciitis

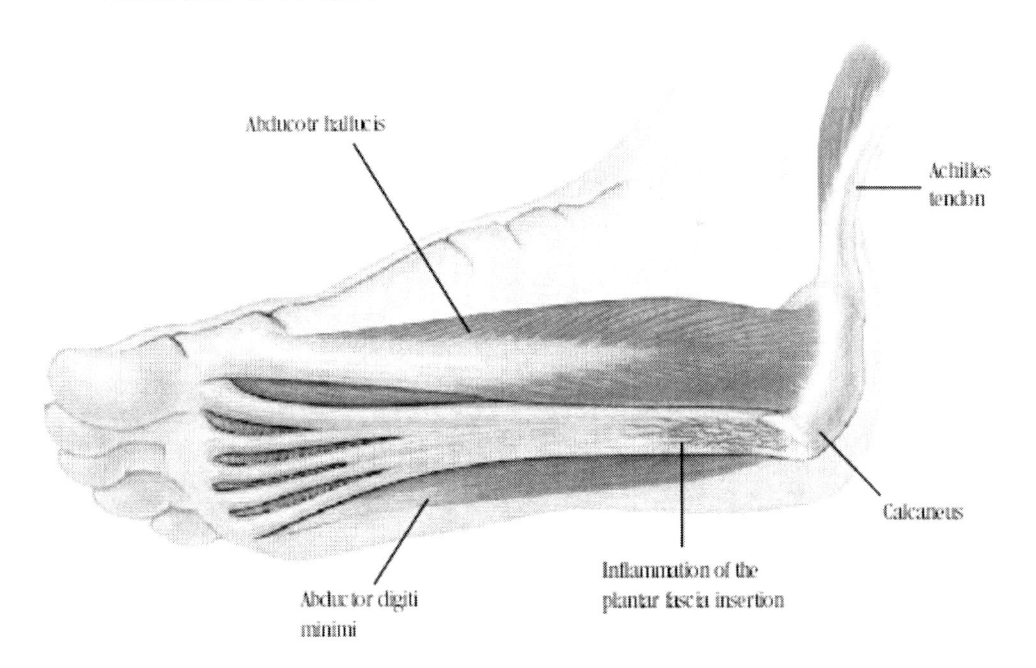

Musculoskeletal Disorders of the Ankles, Feet and Toes
2j) Diagram 40.
Foot Pain

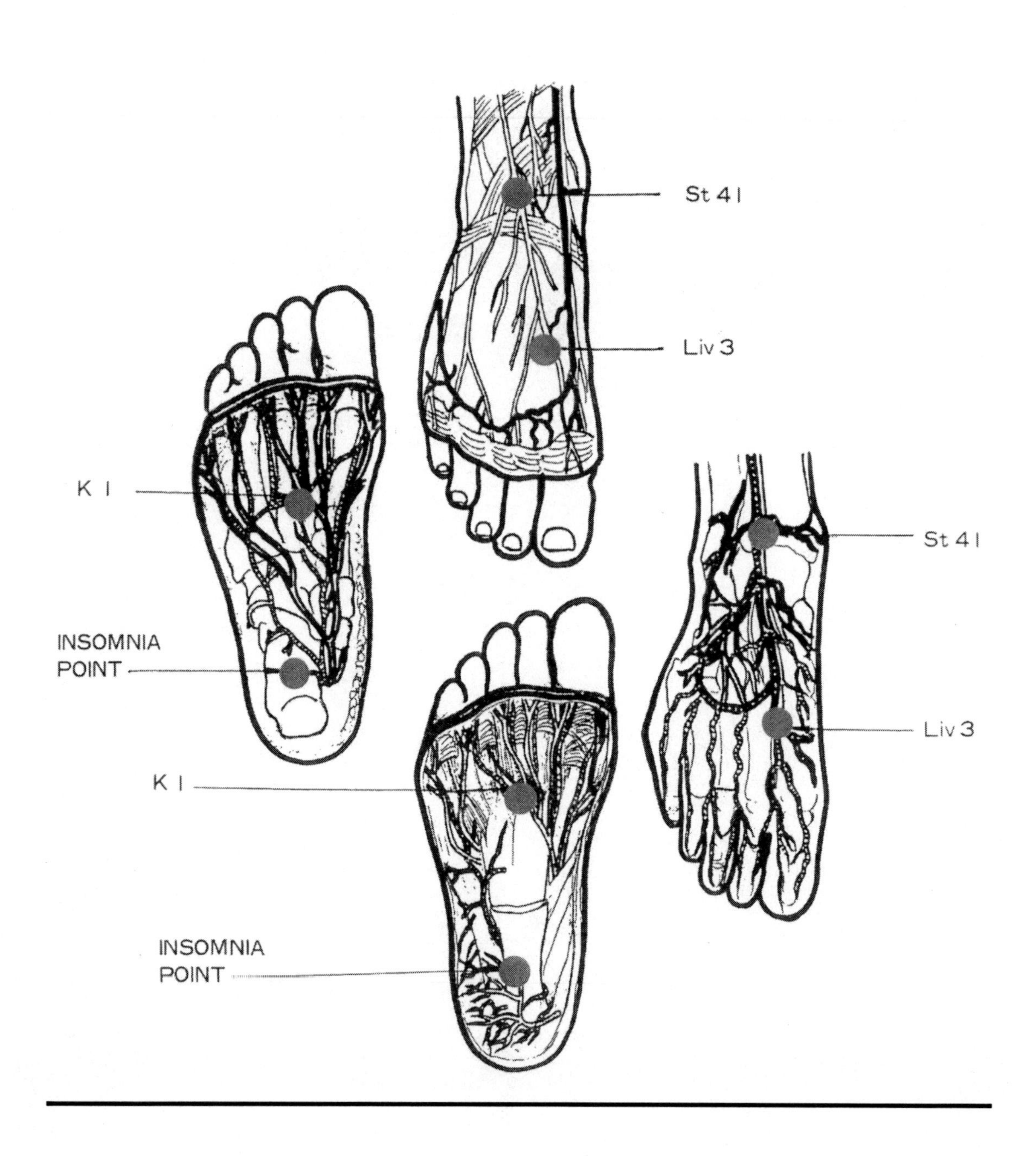

Musculoskeletal Disorders of the Ankles, Feet and Toes
2j) Diagram 41.
Ankle Pain - Osteoarthritis Lateral Ankle Pain – Sprain

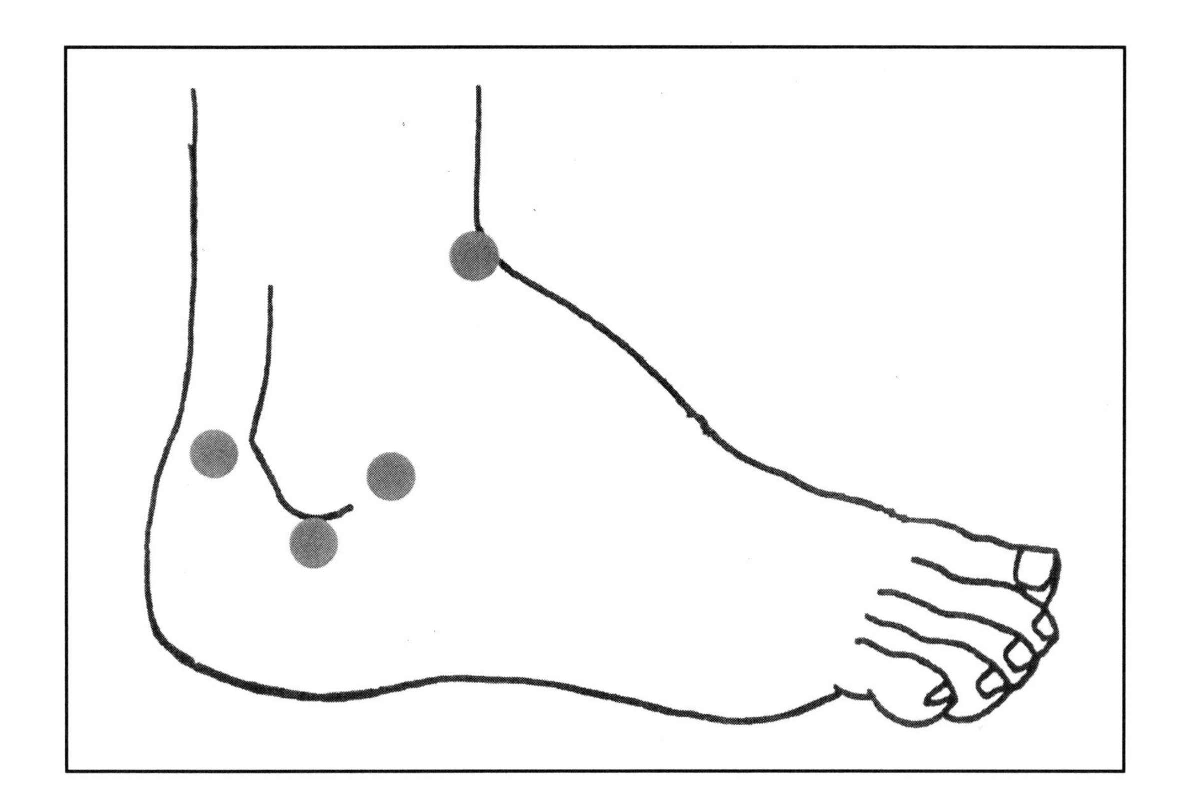

Musculoskeletal Disorders of the Ankles, Feet and Toes
2j) Diagram 42.
Osteoarthritis

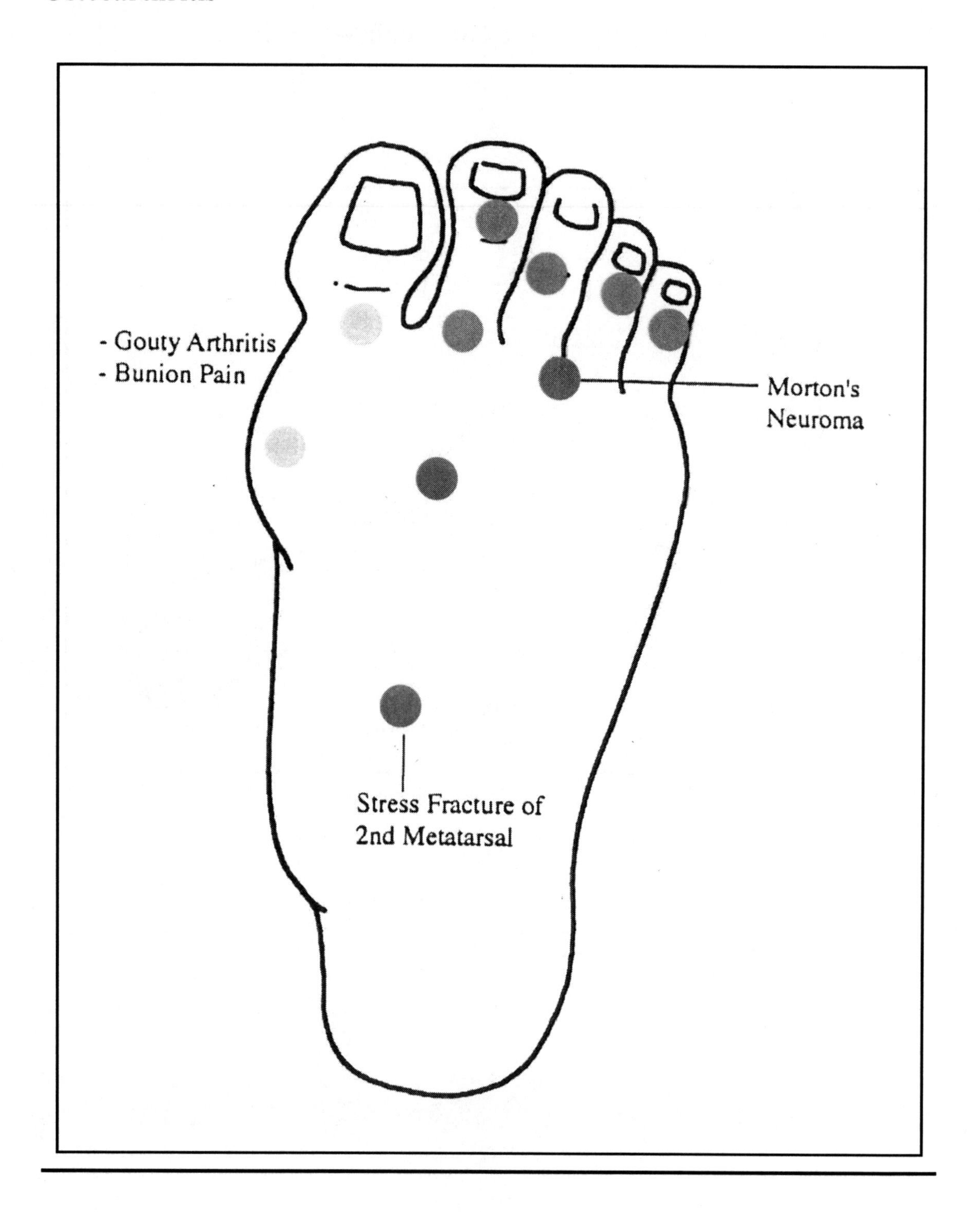

Musculoskeletal Disorders of the Ankles, Feet and Toes
2j) Diagram 43.
Osteoarthritis

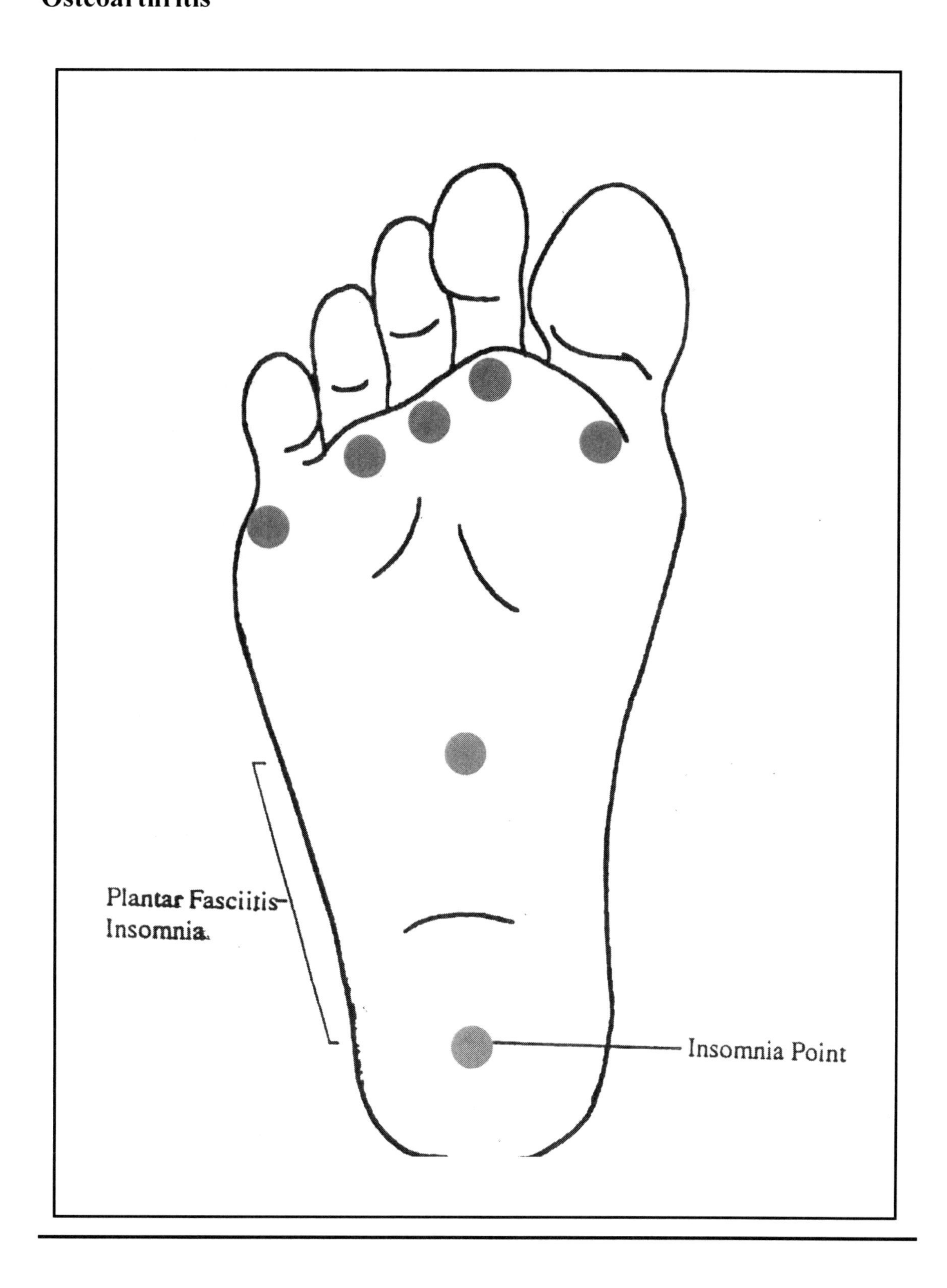

Clinical Treatment of Lupus
Diagram 2k.
Lupus

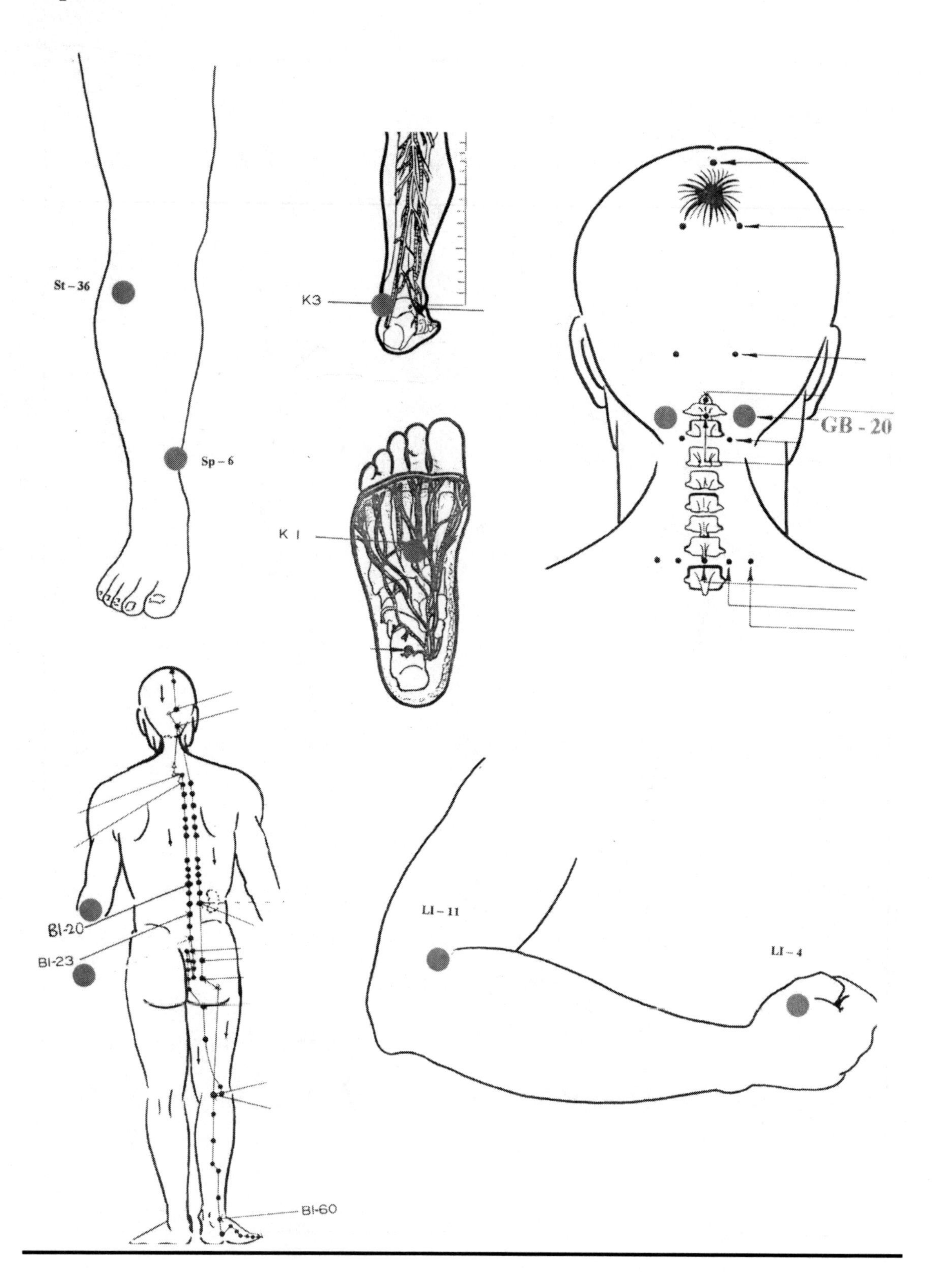

Clinical Treatment of Osteoporosis
Diagram 21.
Spinal Pain - Degenerative Disc Disease, Osteoporosis

(GV)

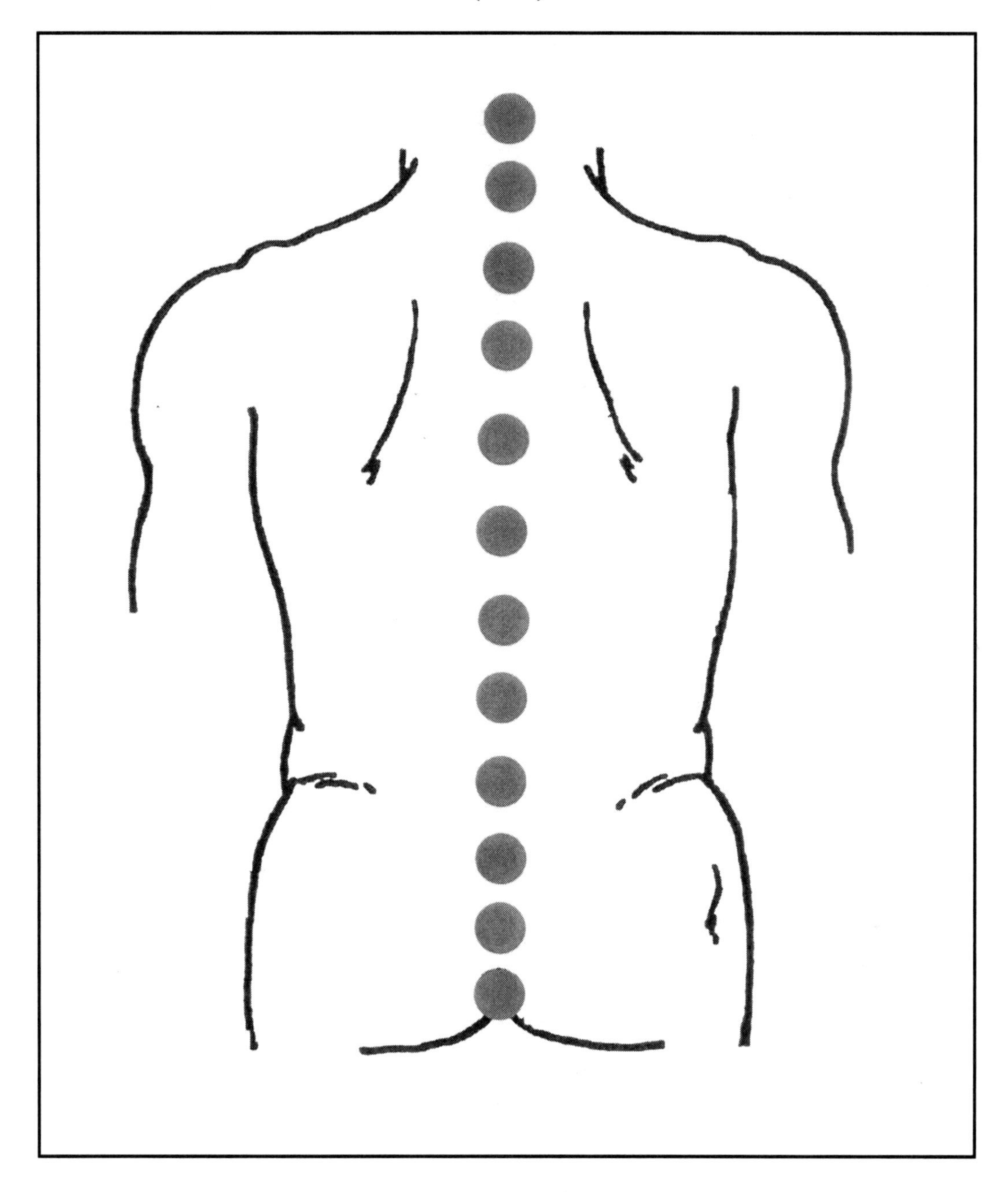

Clinical Treatment of Rheumatoid Arthritis
Diagram 2m.
Rheumatoid Arthritis

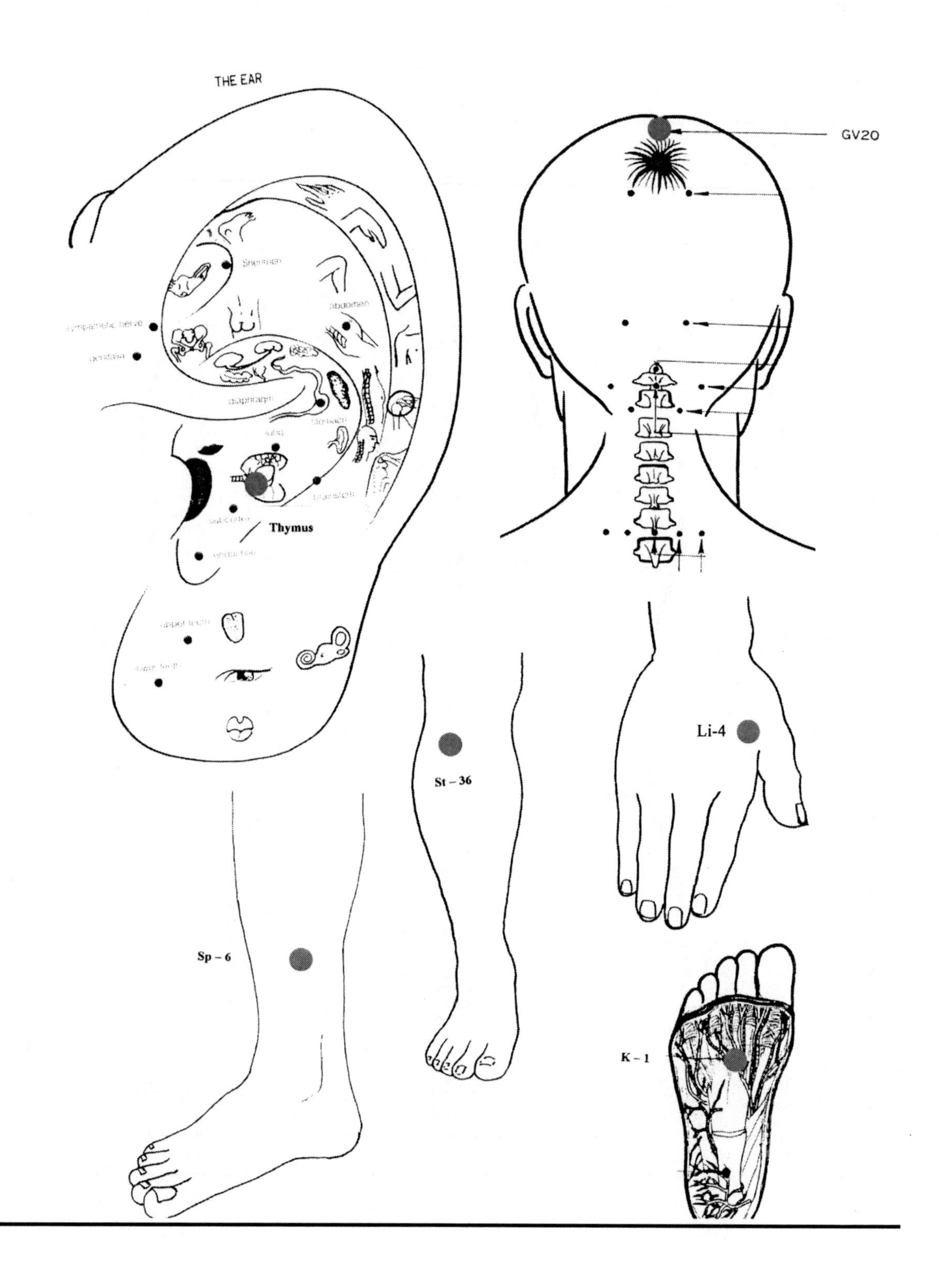

Cardiovascular Disorders
Diagram 3a.
Hypertension

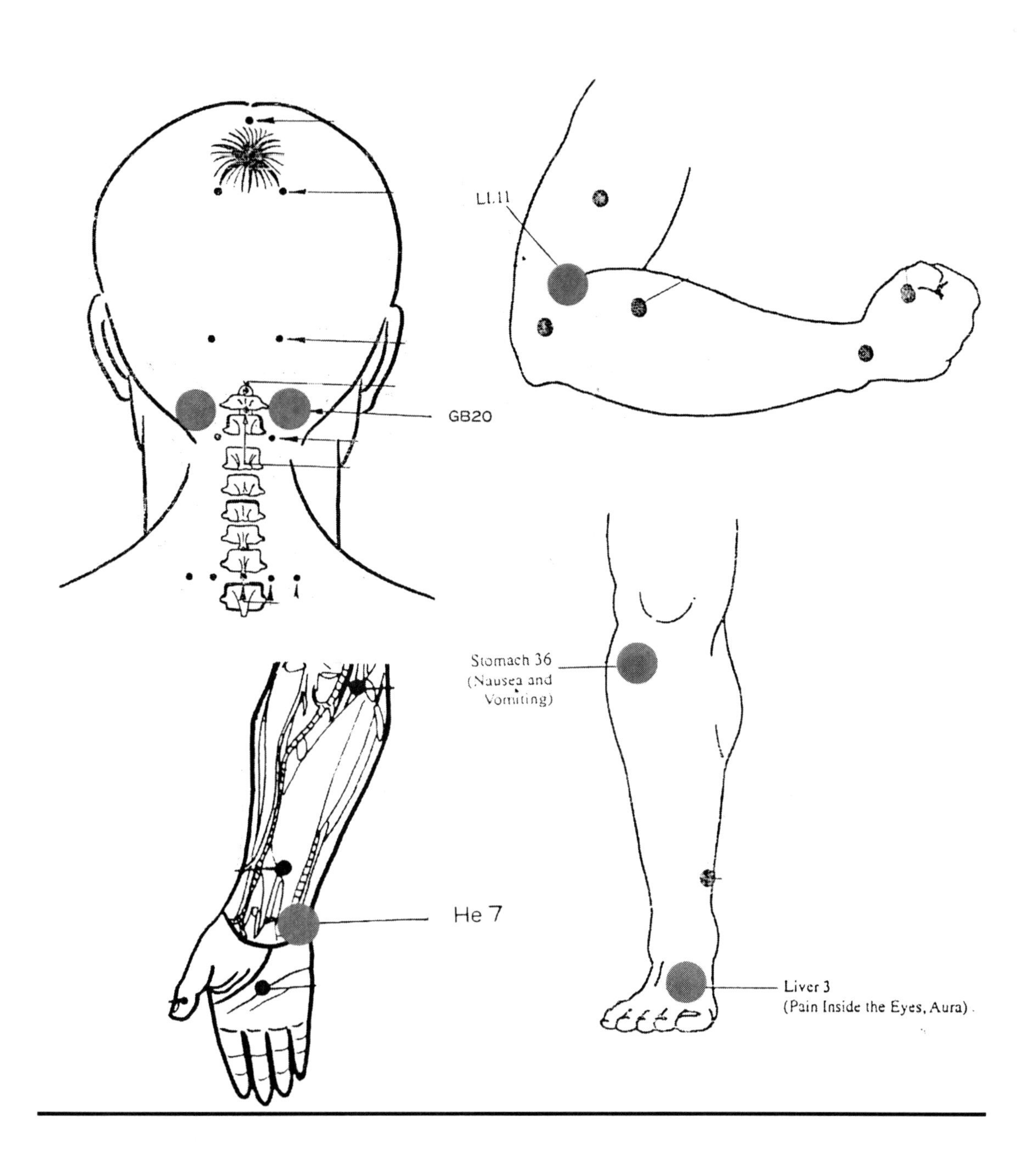

Cardiovascular Disorders
Diagram 3b.
Hypotension

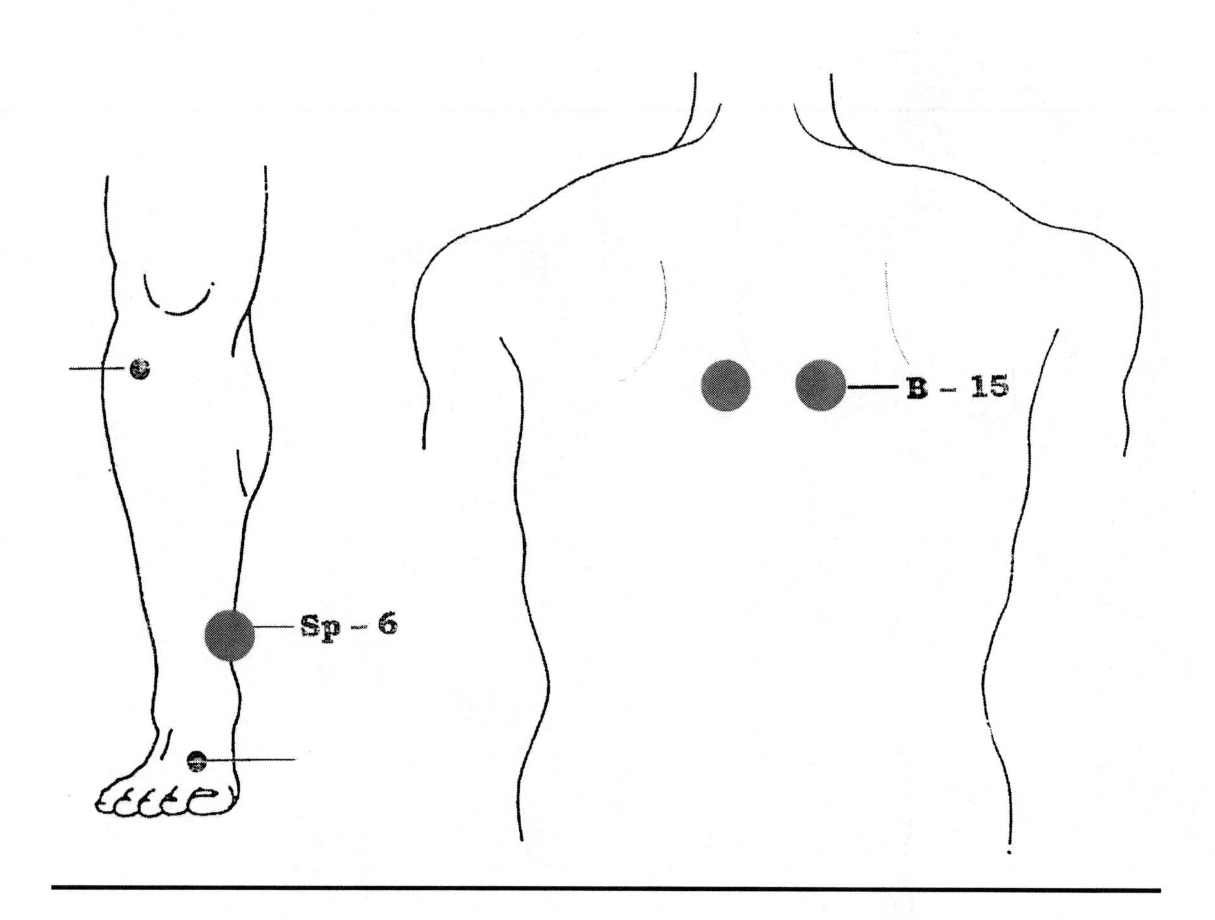

Gastroenterology
Diagram 4a.
Anorexia

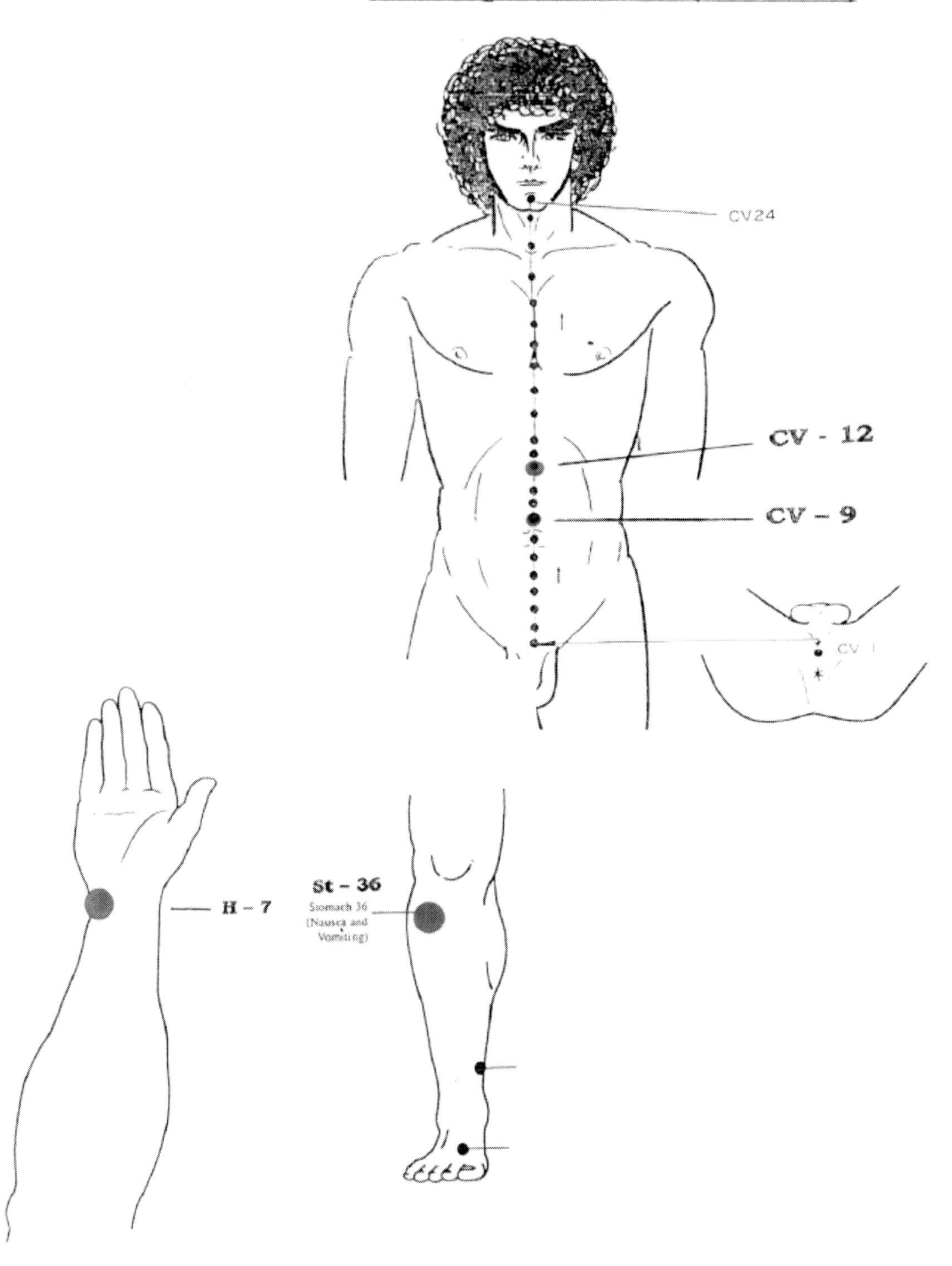

Gastroenterology
Diagram 4b.
Colitis, Enterocolitis and Crohn's disease

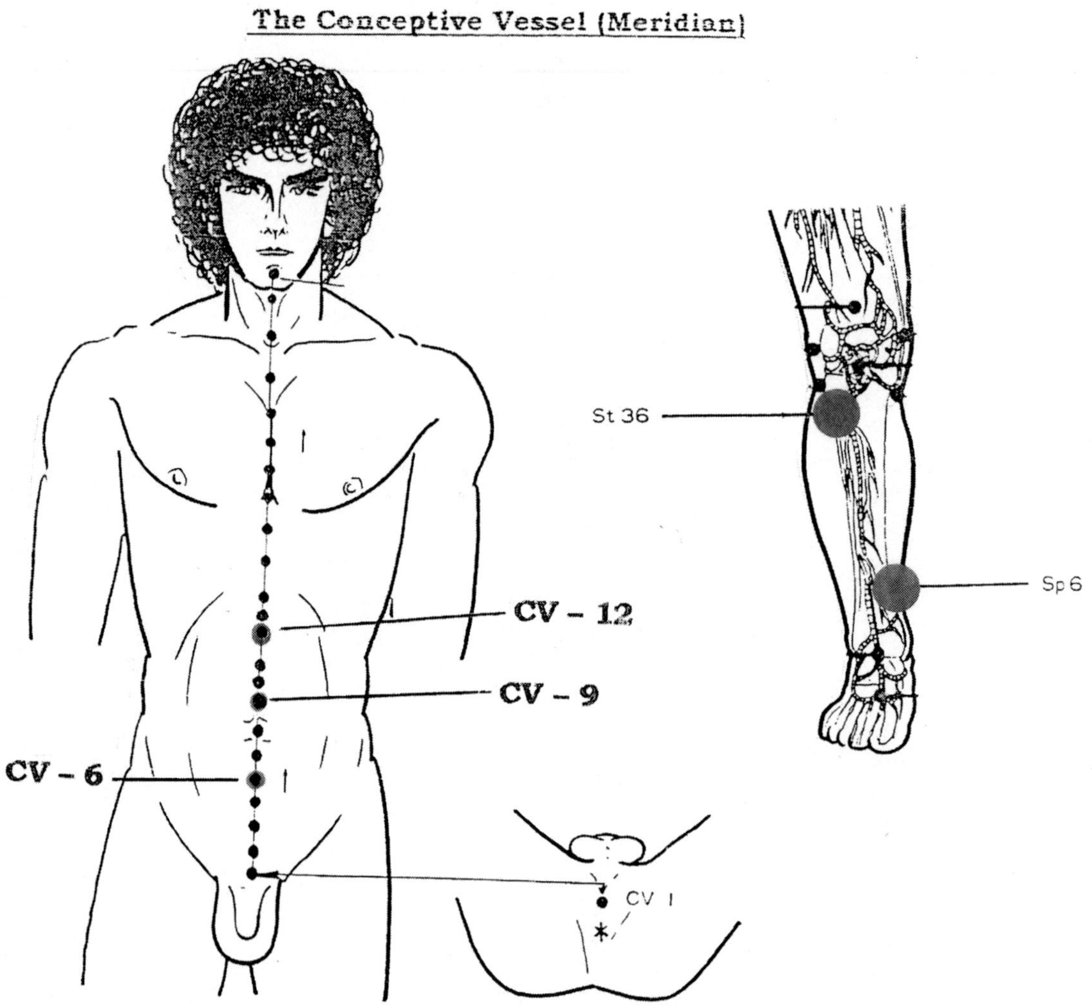

Gastroenterology
Diagram 4c.
Diarrhea & Constipation

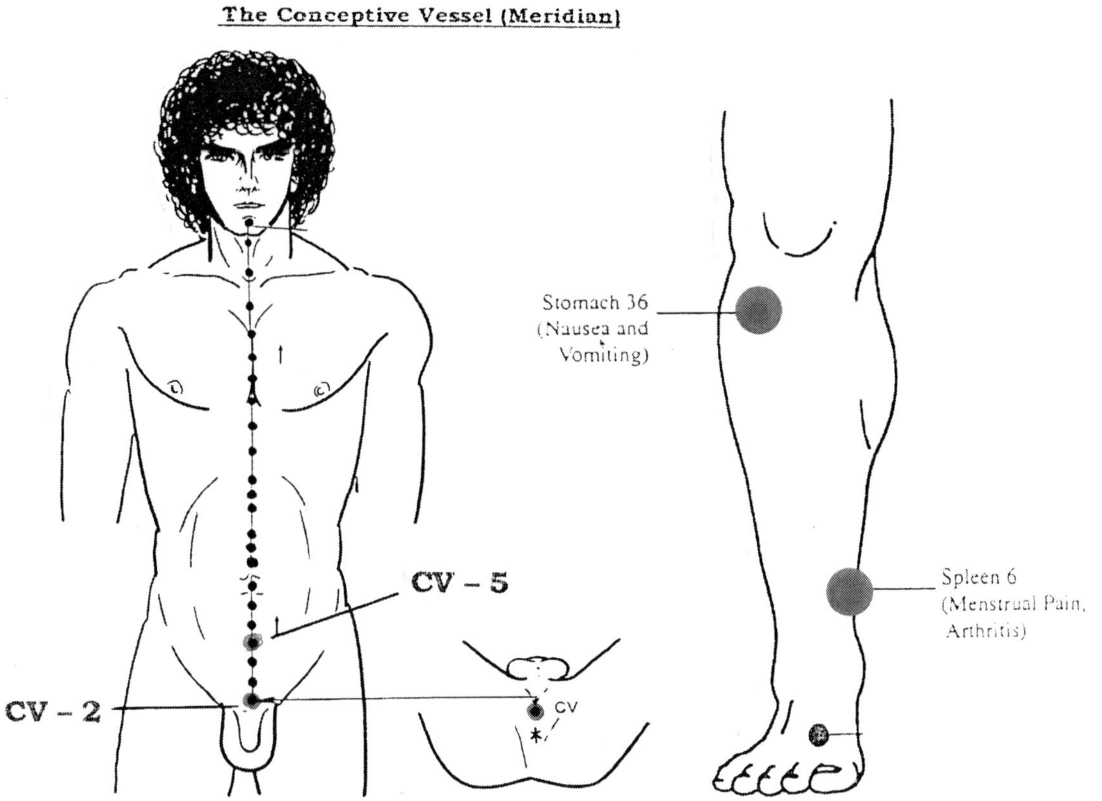

The Conceptive Vessel (Meridian)

CV - 5

CV - 2

CV

Stomach 36
(Nausea and
Vomiting)

Spleen 6
(Menstrual Pain,
Arthritis)

Gastroenterology
Diagram 4d.
Hepatitis

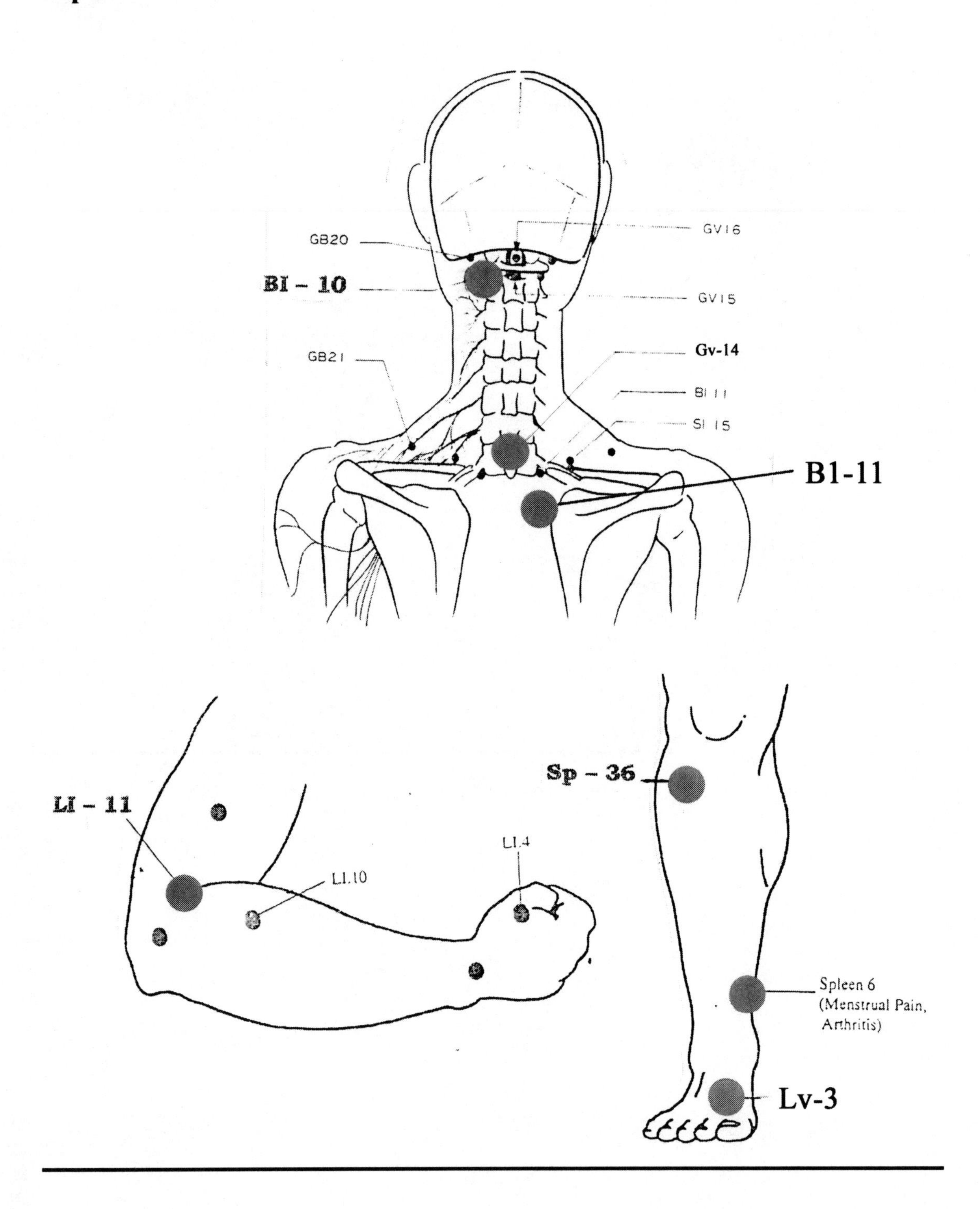

Gastroenterology
Diagram 4e.
Ileitis & Paralytic Ileus

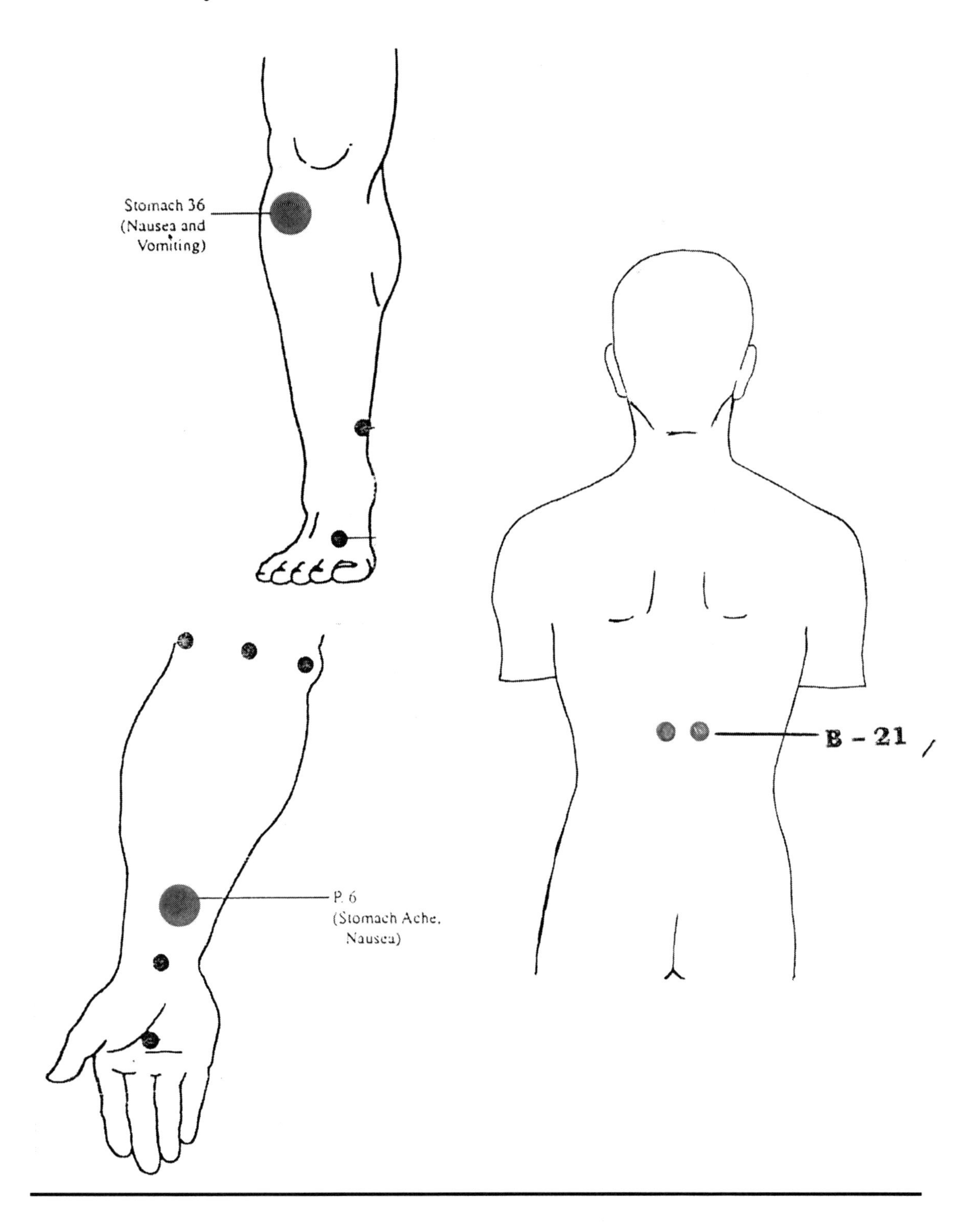

Gastroenterology
Diagram 4f.
Abdominal Pain/ Irritable Bowel Syndrome

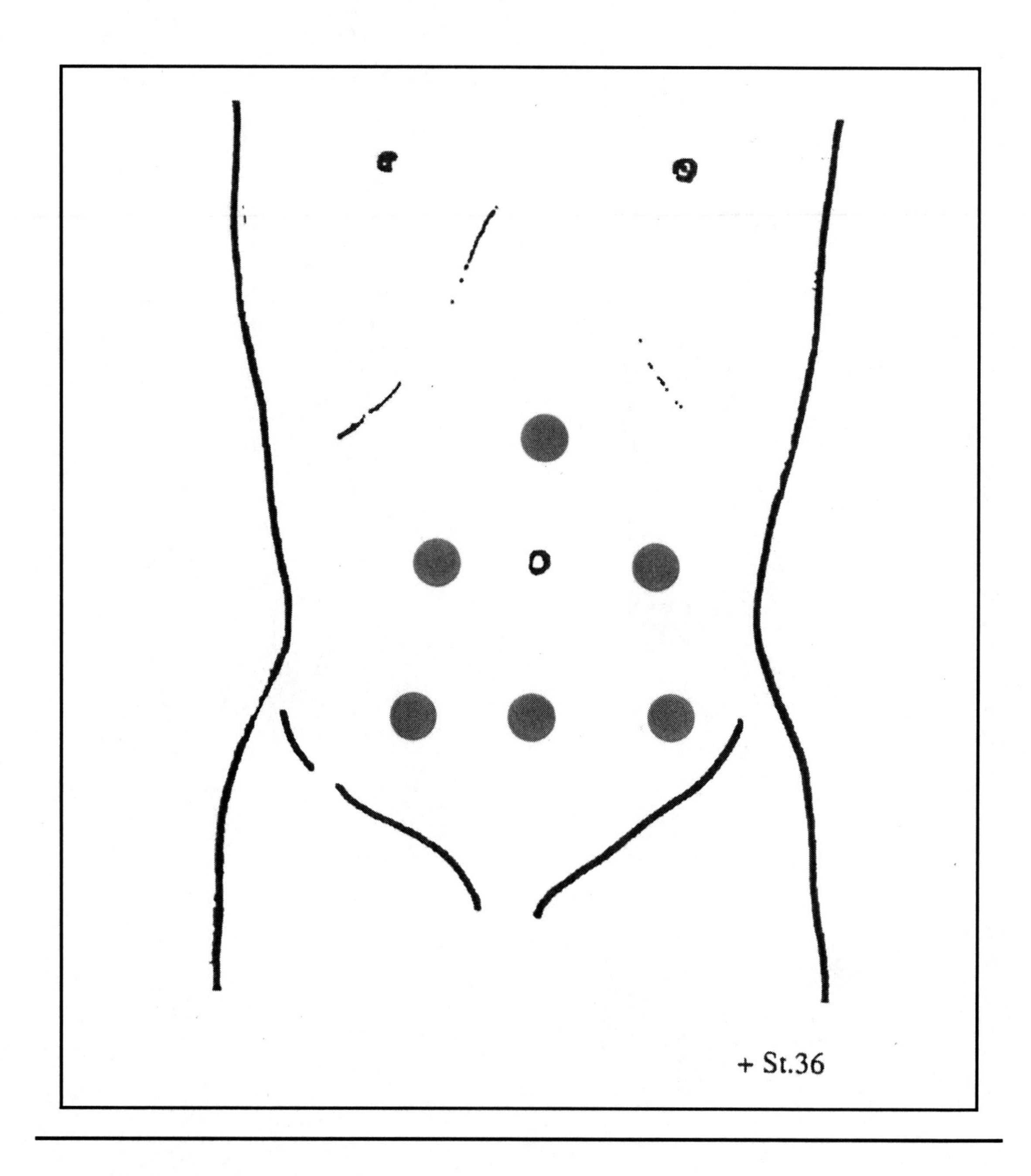

Gastroenterology
Diagram 4g.
Obesity

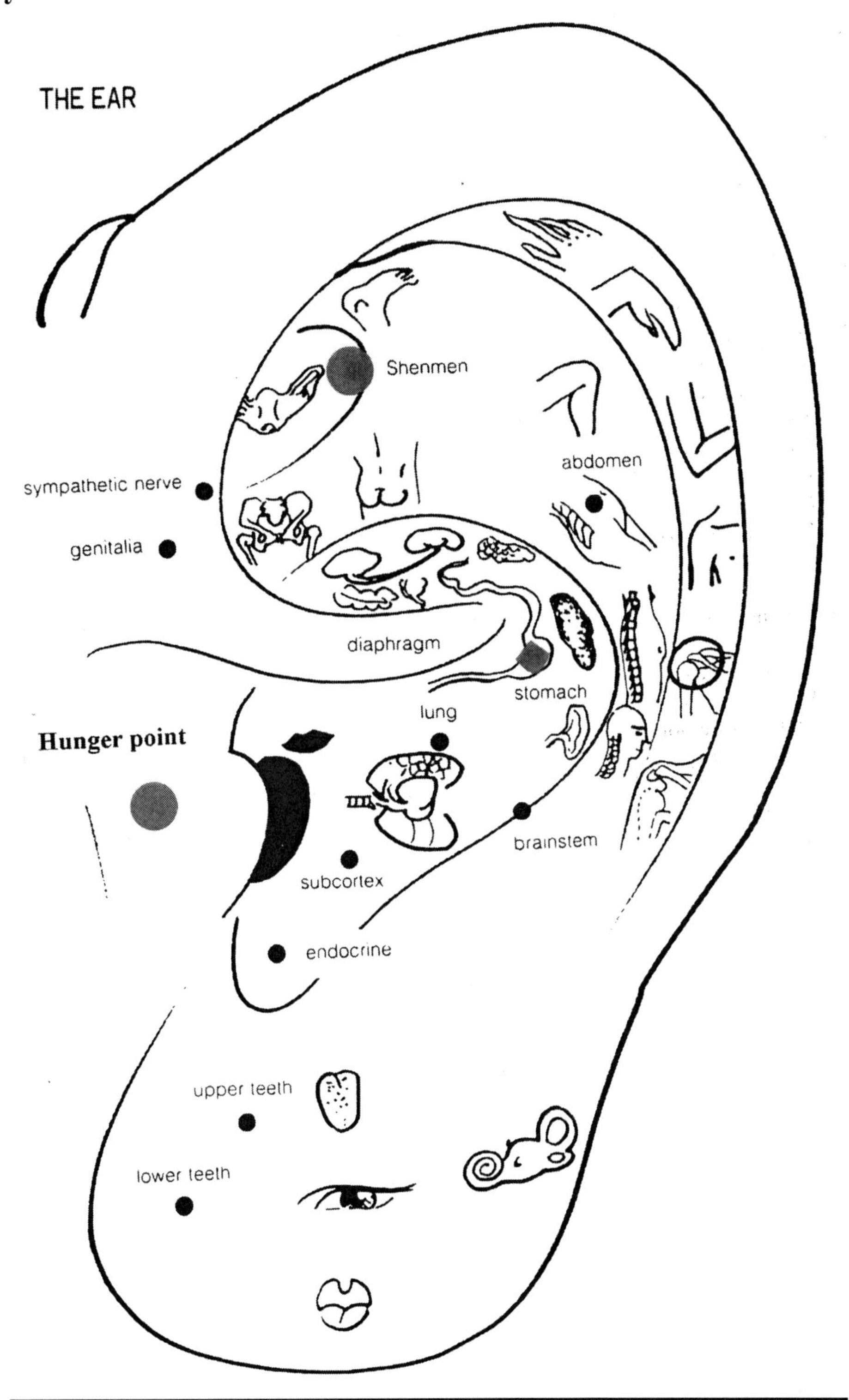

Genitourology
Diagram 5a.
Impotence

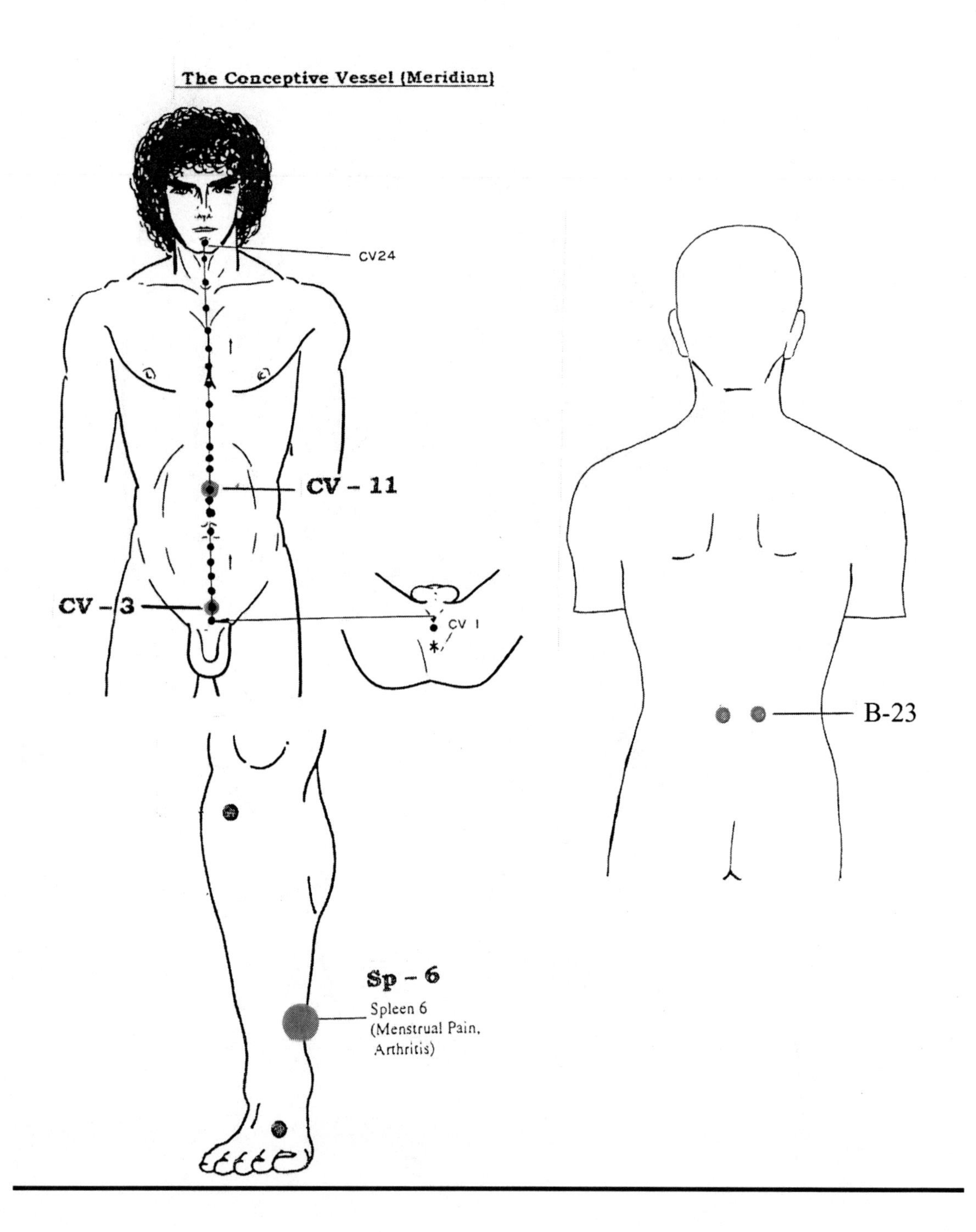

Genitourology
Diagram 5b.
Nephritis

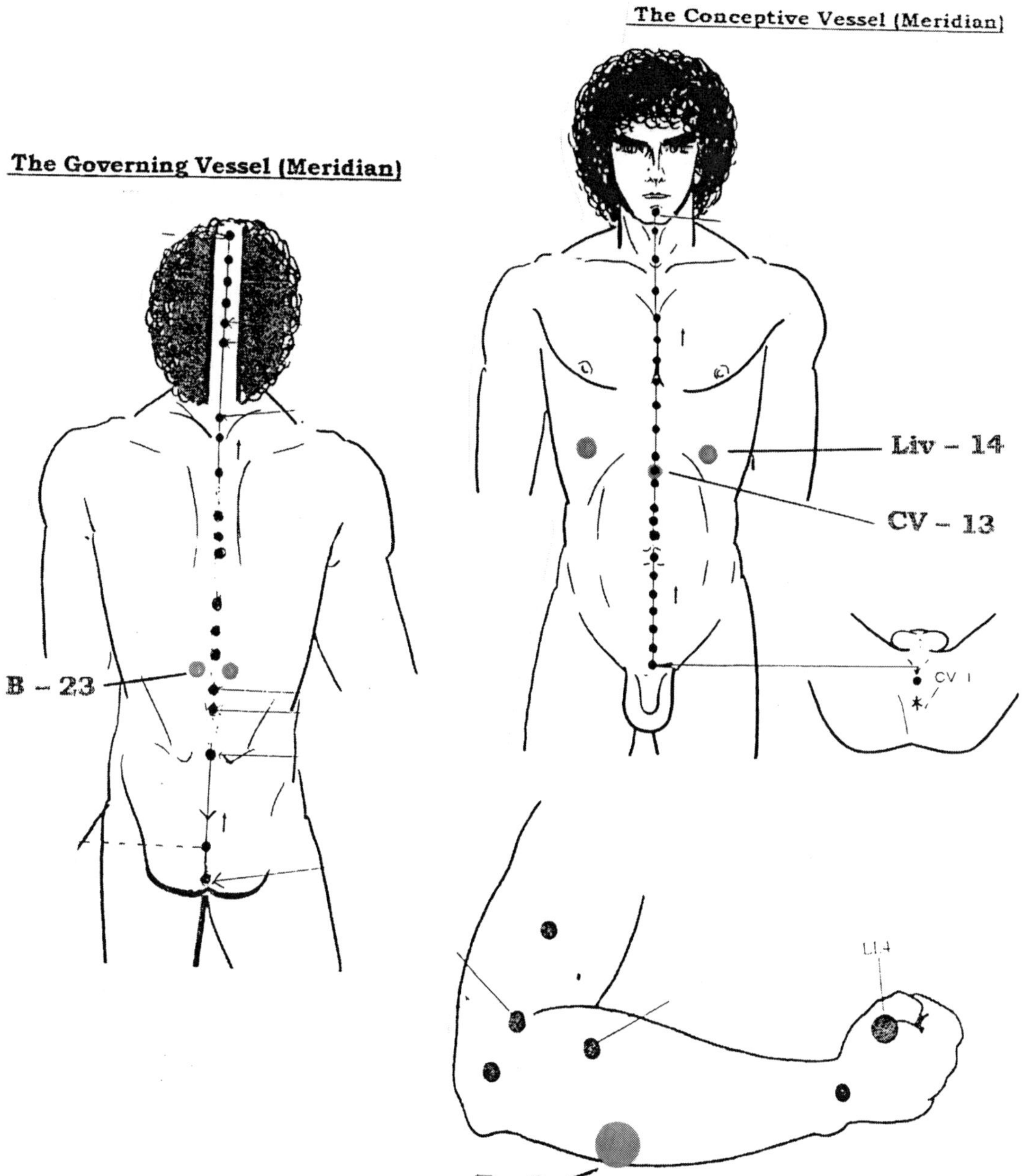

Genitourology
Diagram 5c.
Prostatitis & Prostate Hypertrophy

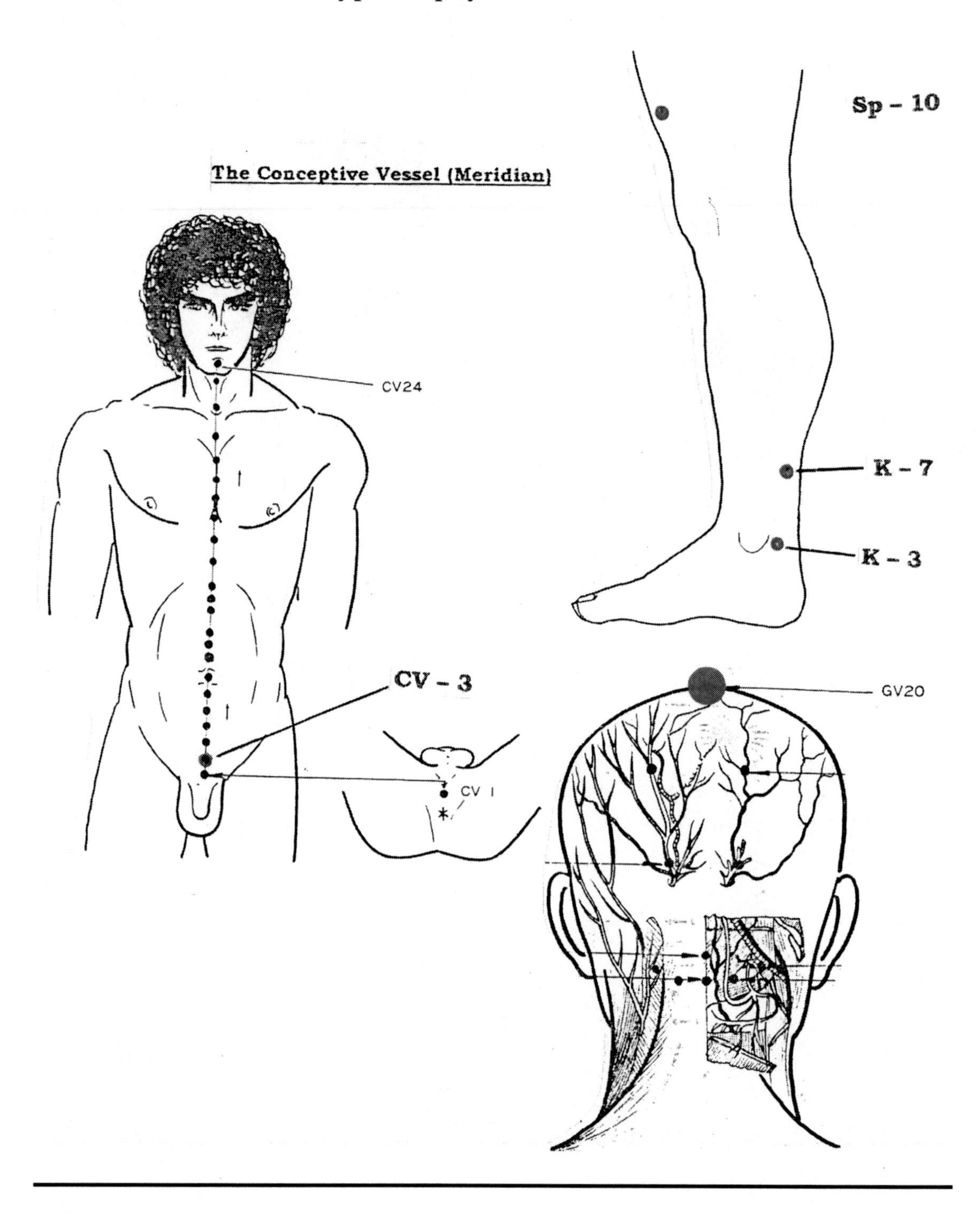

Genitourology
Diagram 5d.
Renal Colic (Kidney Stone)

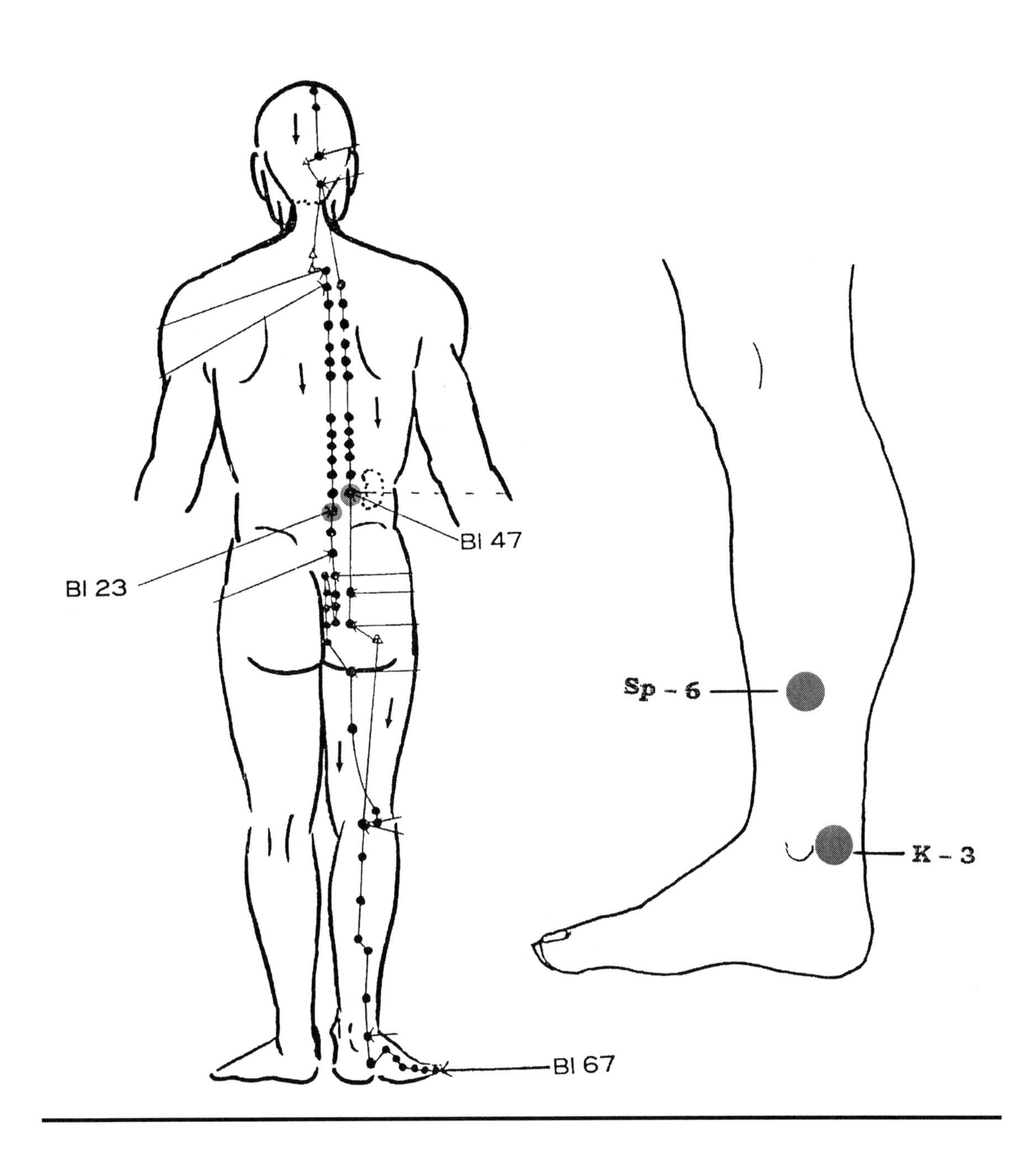

Immune System
Diagram 6.
Immune System Problems – e.g., Asthma, AIDS, eczema, hay fever, lupus, rheumatoid arthritis, psoriasis, Crohn's disease, scleroderma

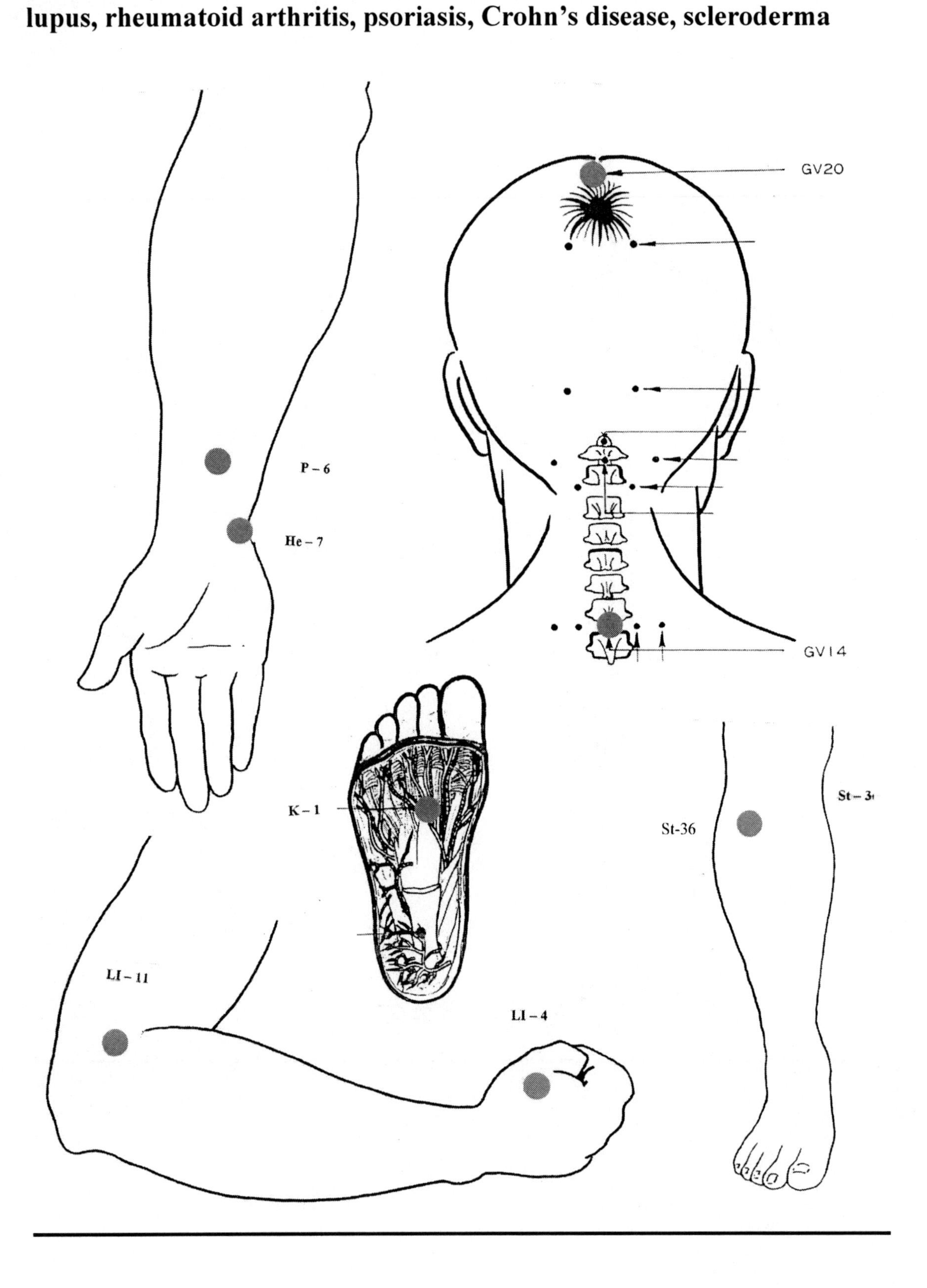

Neurological Problems
Diagram 7a.
Alzheimer's Disease

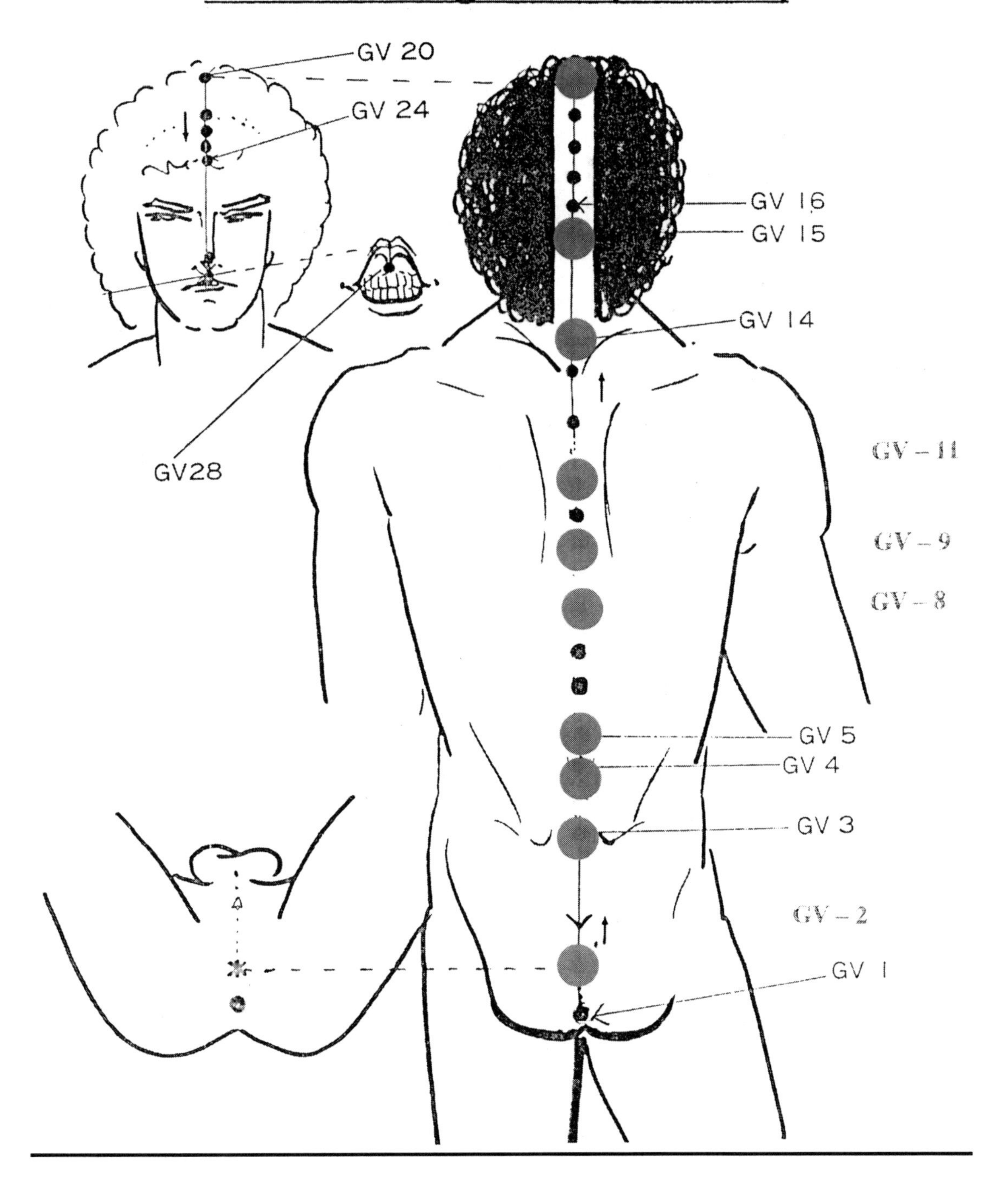

Neurological Problems
Diagram 7b.
Restless Leg Syndrome - Diabetic Neuropathy, Ischemic Leg Pain

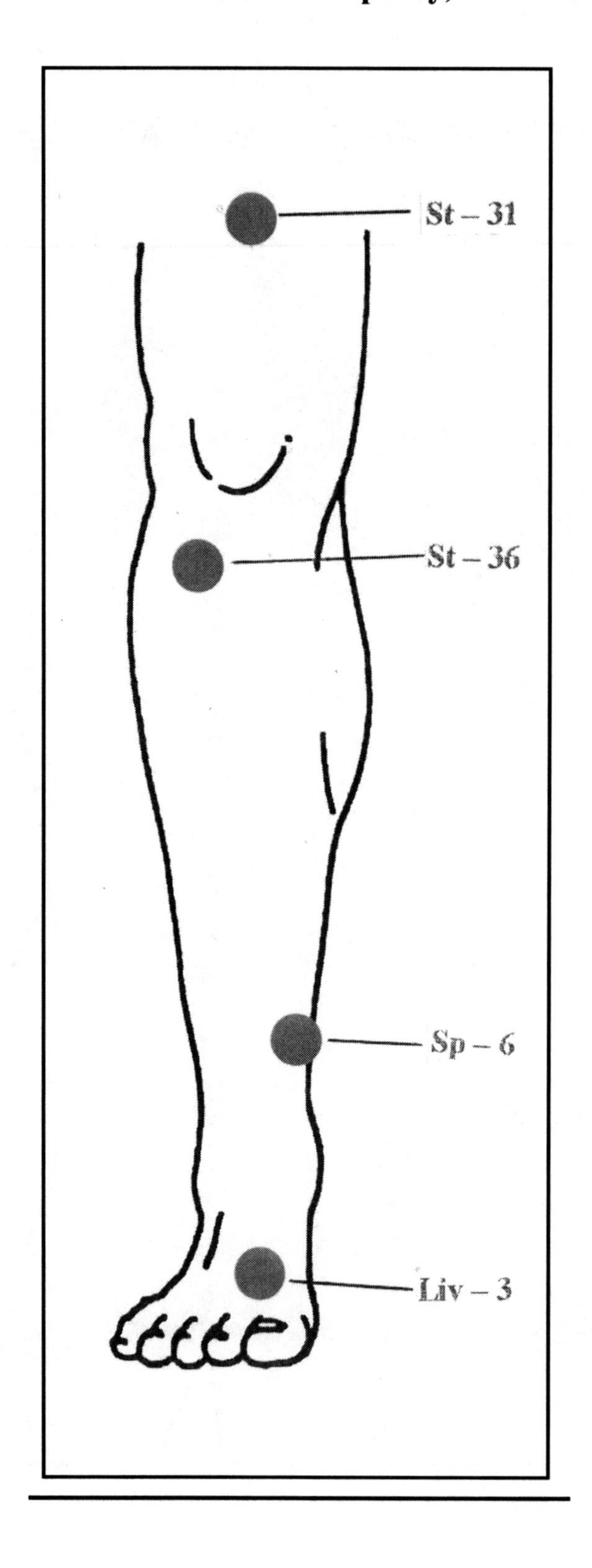

Neurological Problems
Diagram 7c.
Insomnia

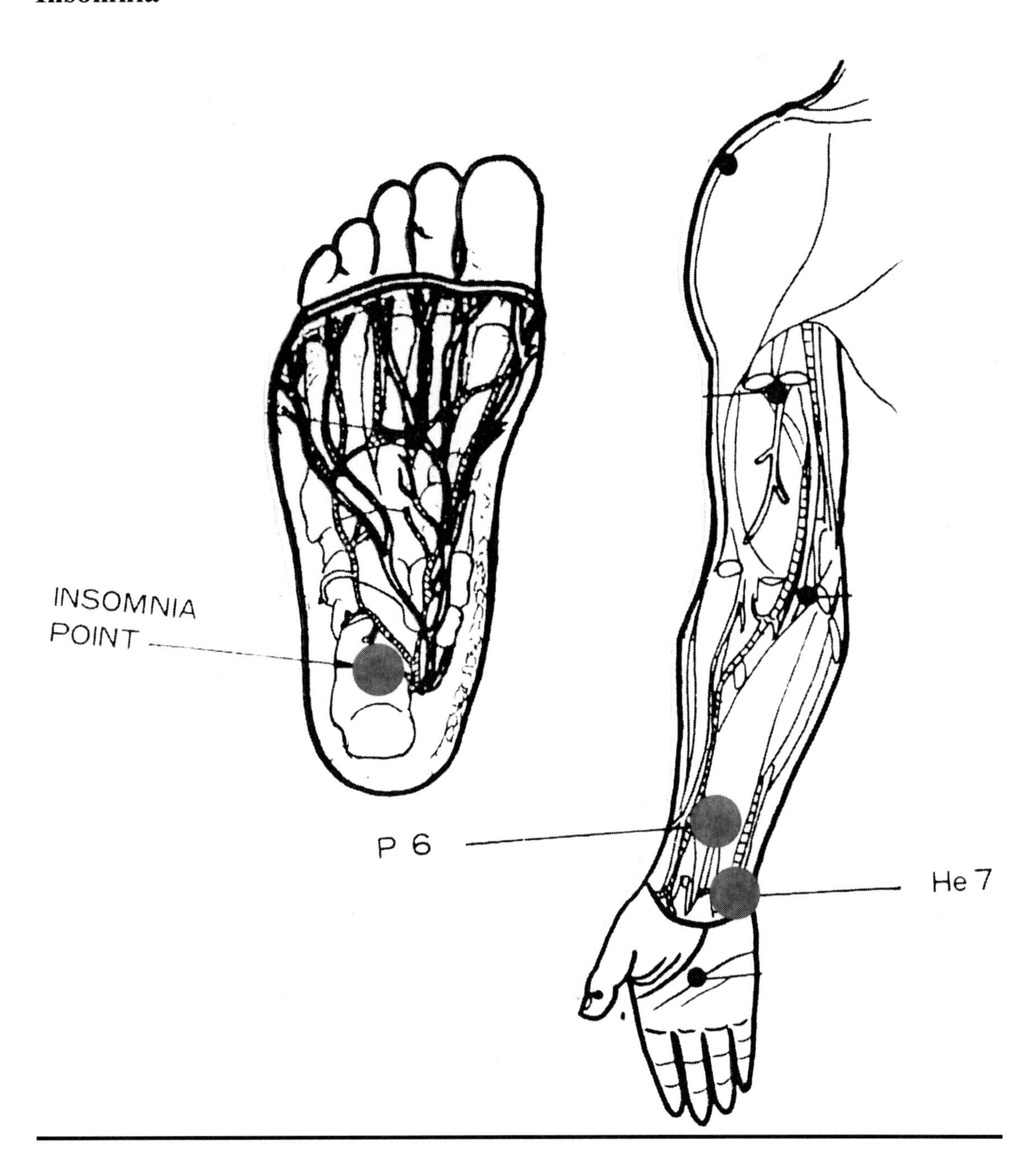

Neurological Problems
Diagram 7d.
Meniere's Disease

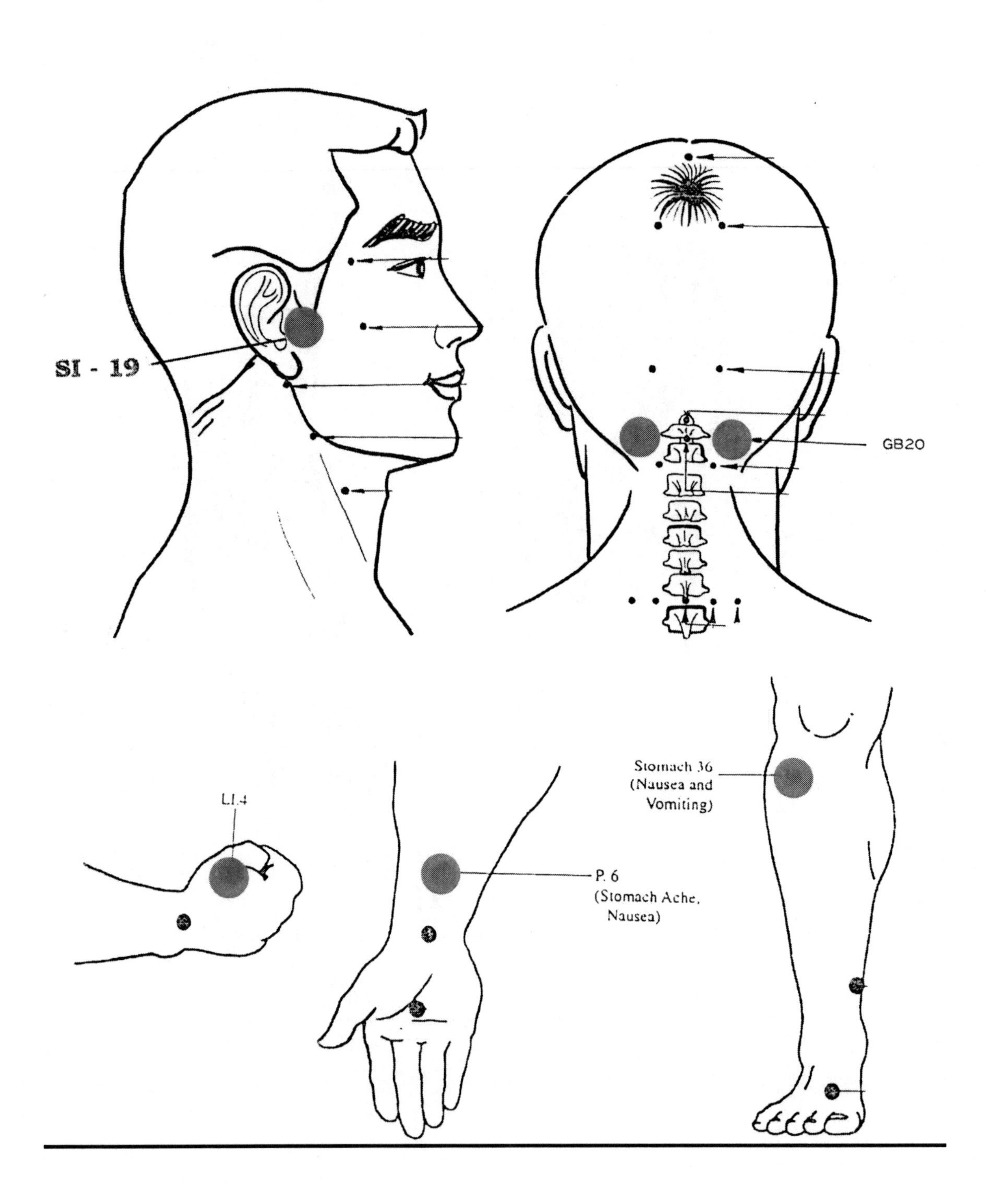

Neurological Problems
Diagram 7e.
Minor Depression and Nervousness

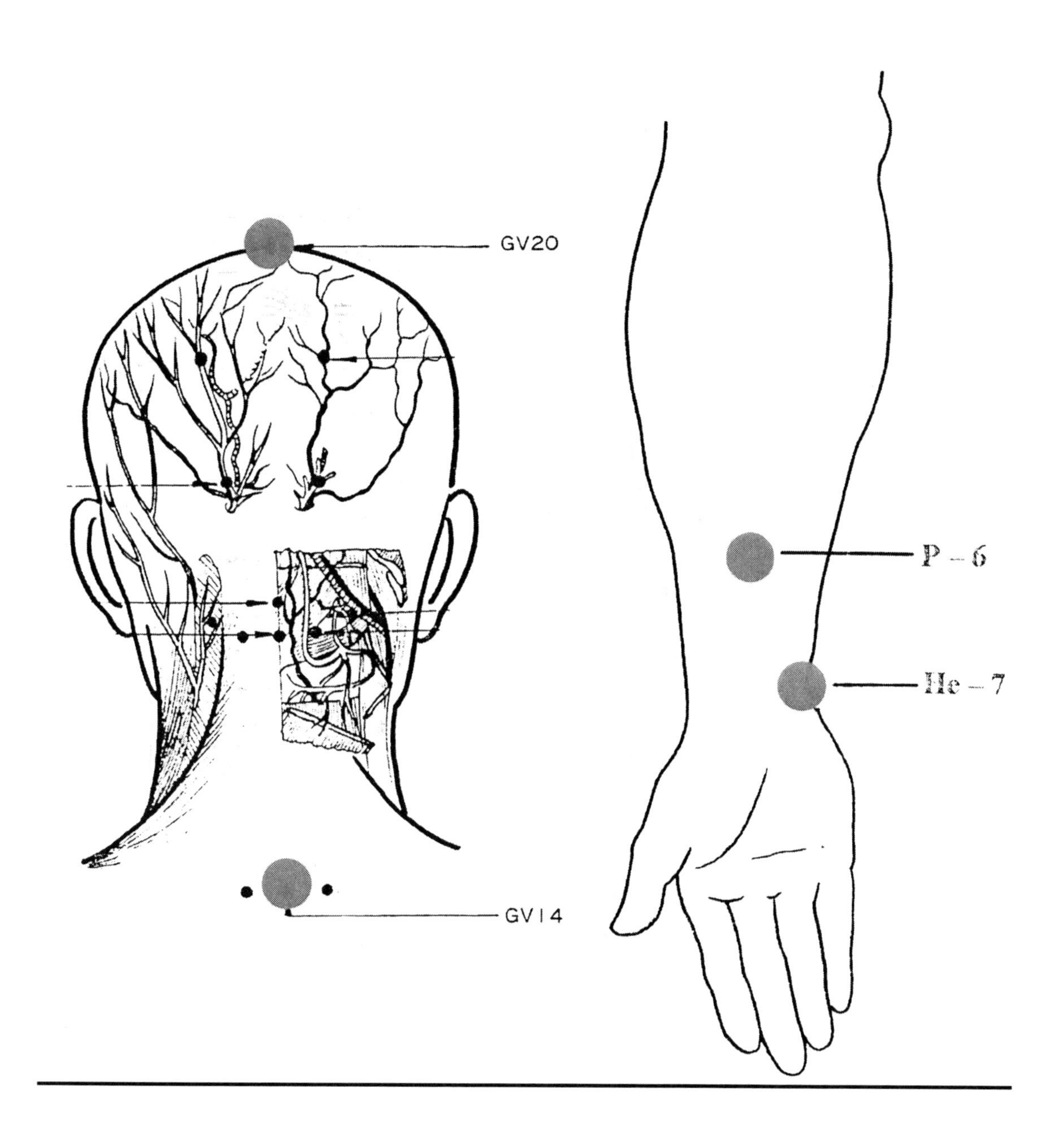

Neurological Problems
Diagram 7j.
Tinnitus

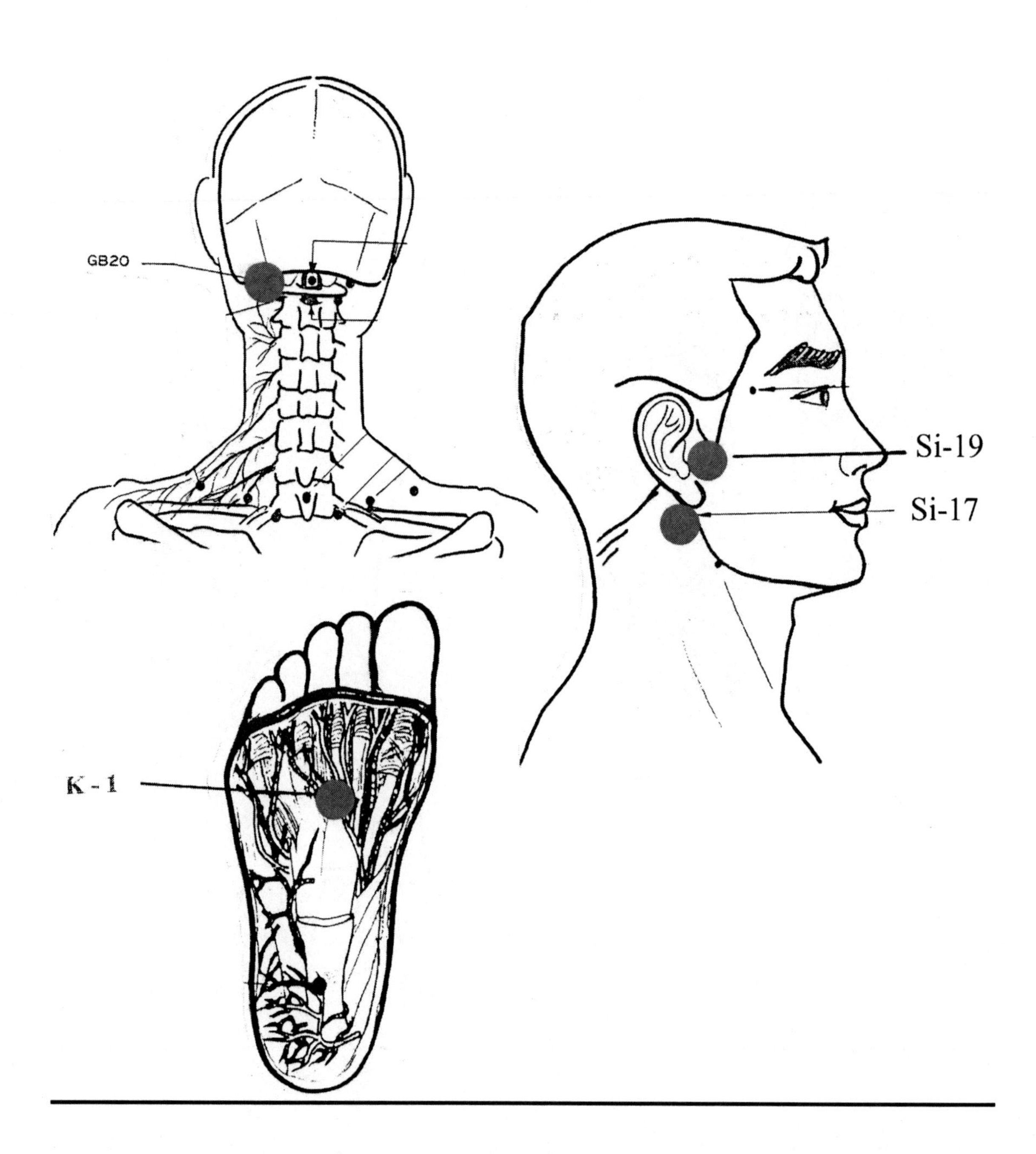

Neurological Problems
Diagram 7k.
Trigeminal Neuralgia

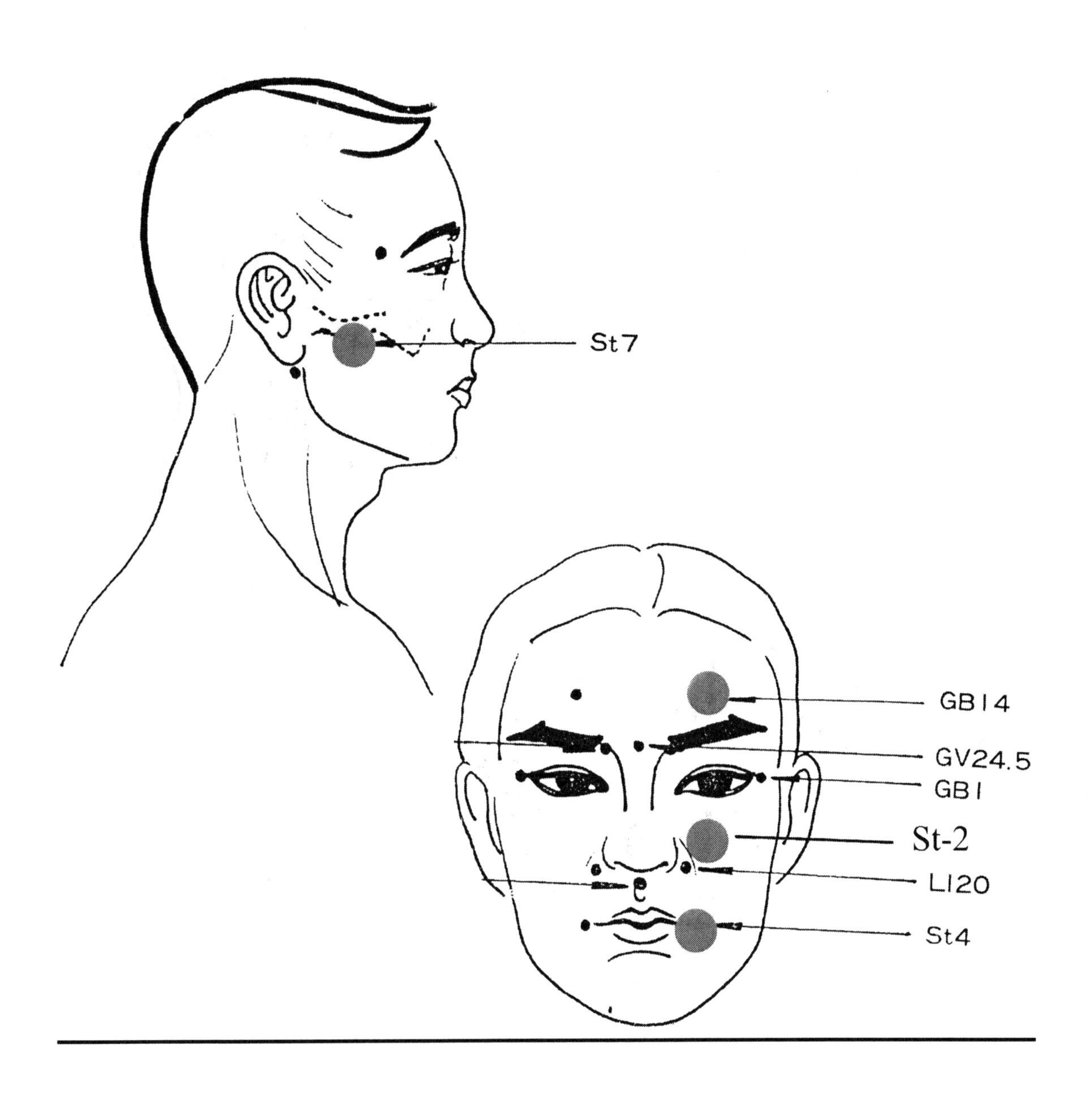

Obstetrics & Gynecology
Diagram 8a.
Fibroids

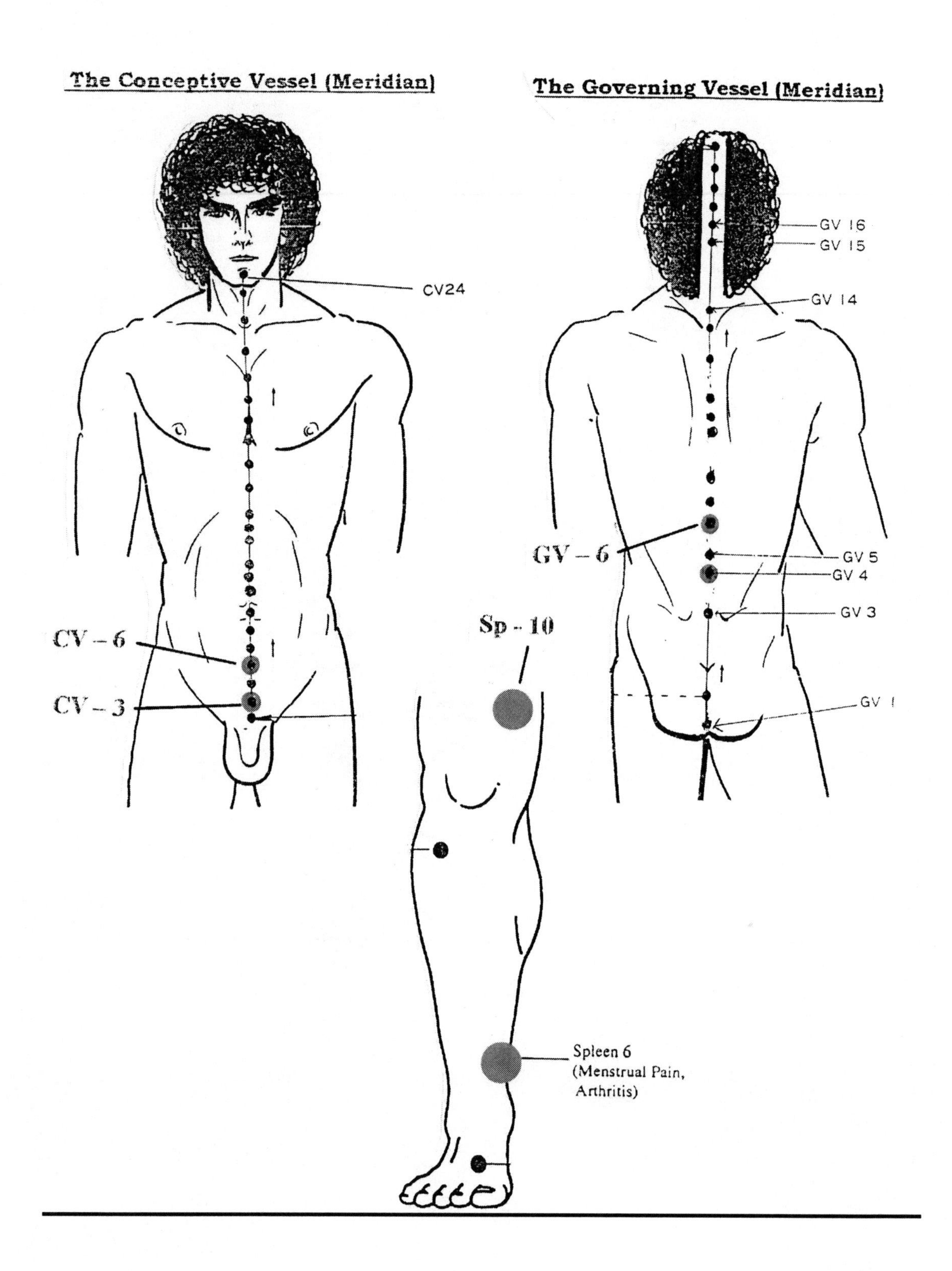

Obstetrics & Gynecology
Diagram 8b.
Menopause

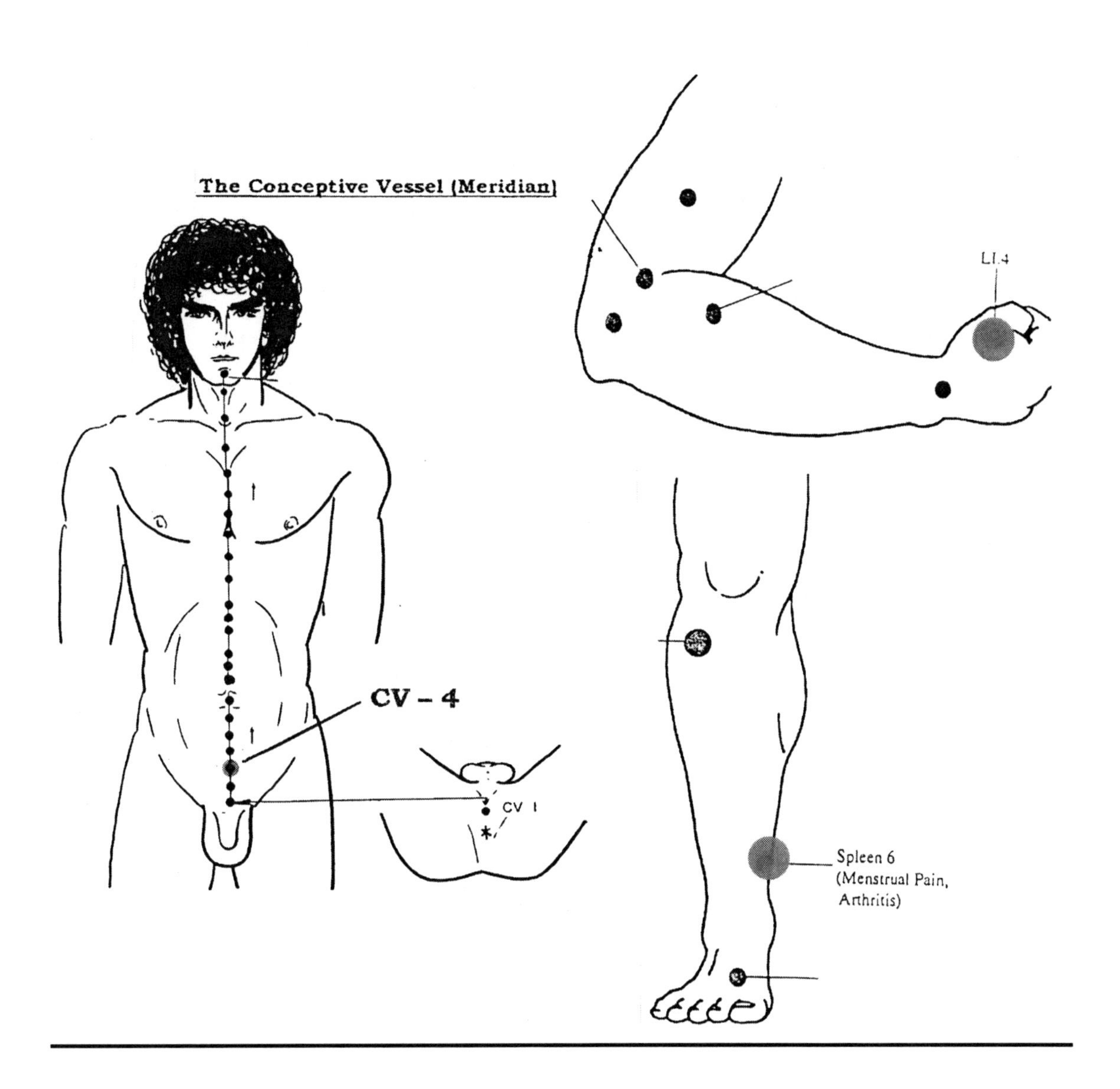

Pediatric/ Children's Problems
Diagram 9b.
Hyperactive Learning/ Hearing Problems - The Ear Points

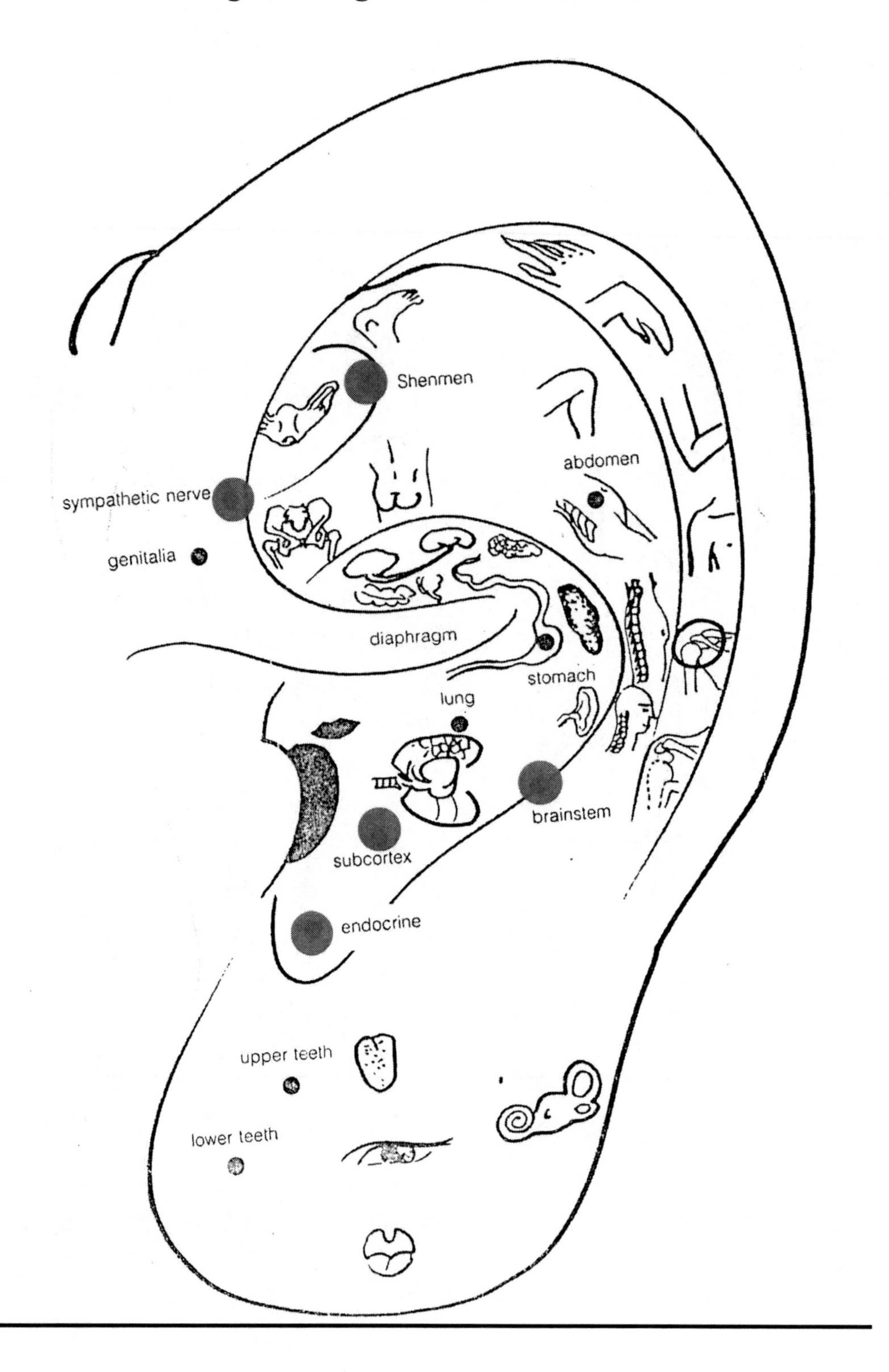

Respiratory System
Diagram 10a.
Allergic Sinusitis

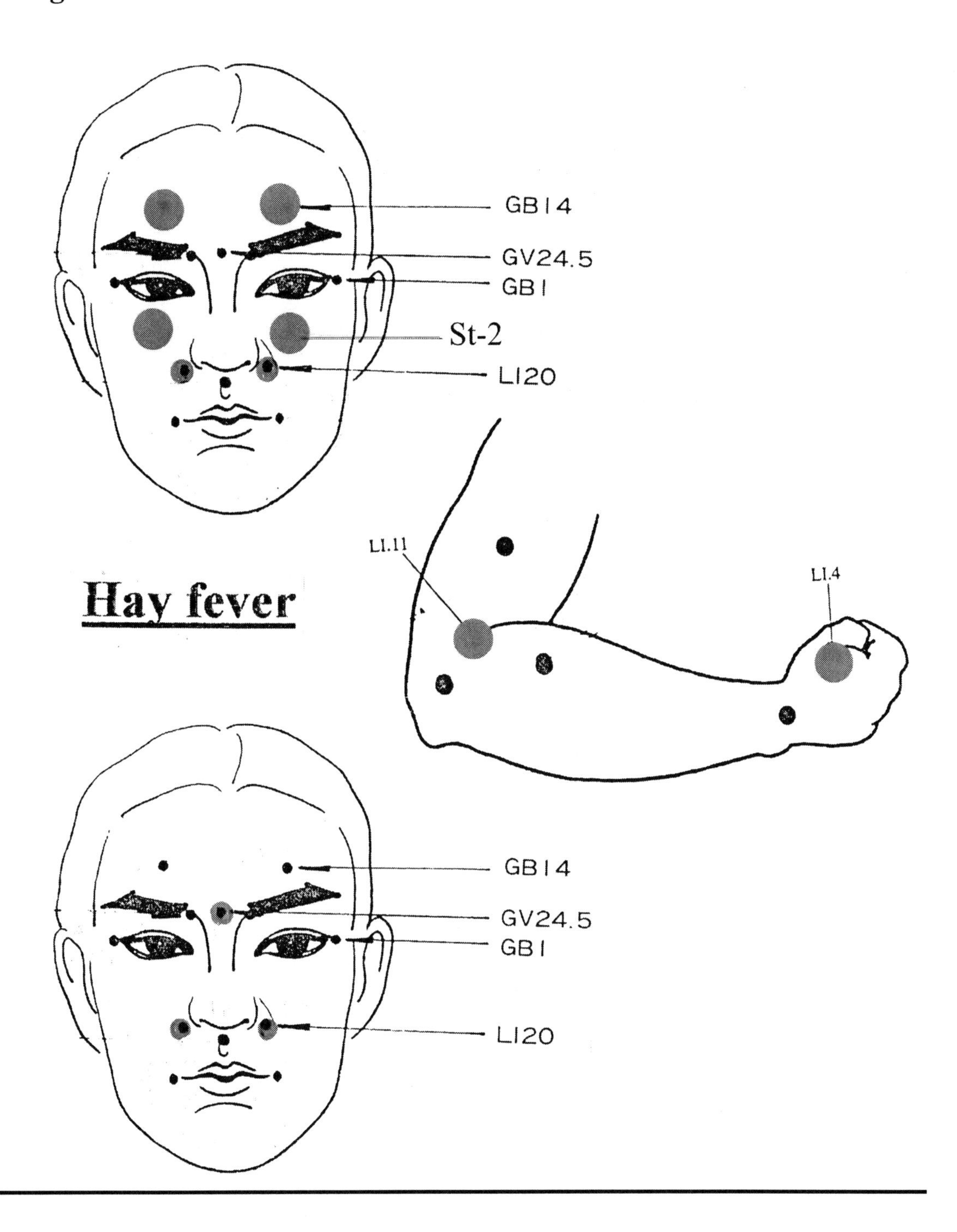

Respiratory System
Diagram 10b.
Asthma, COPD, Chronic Cough

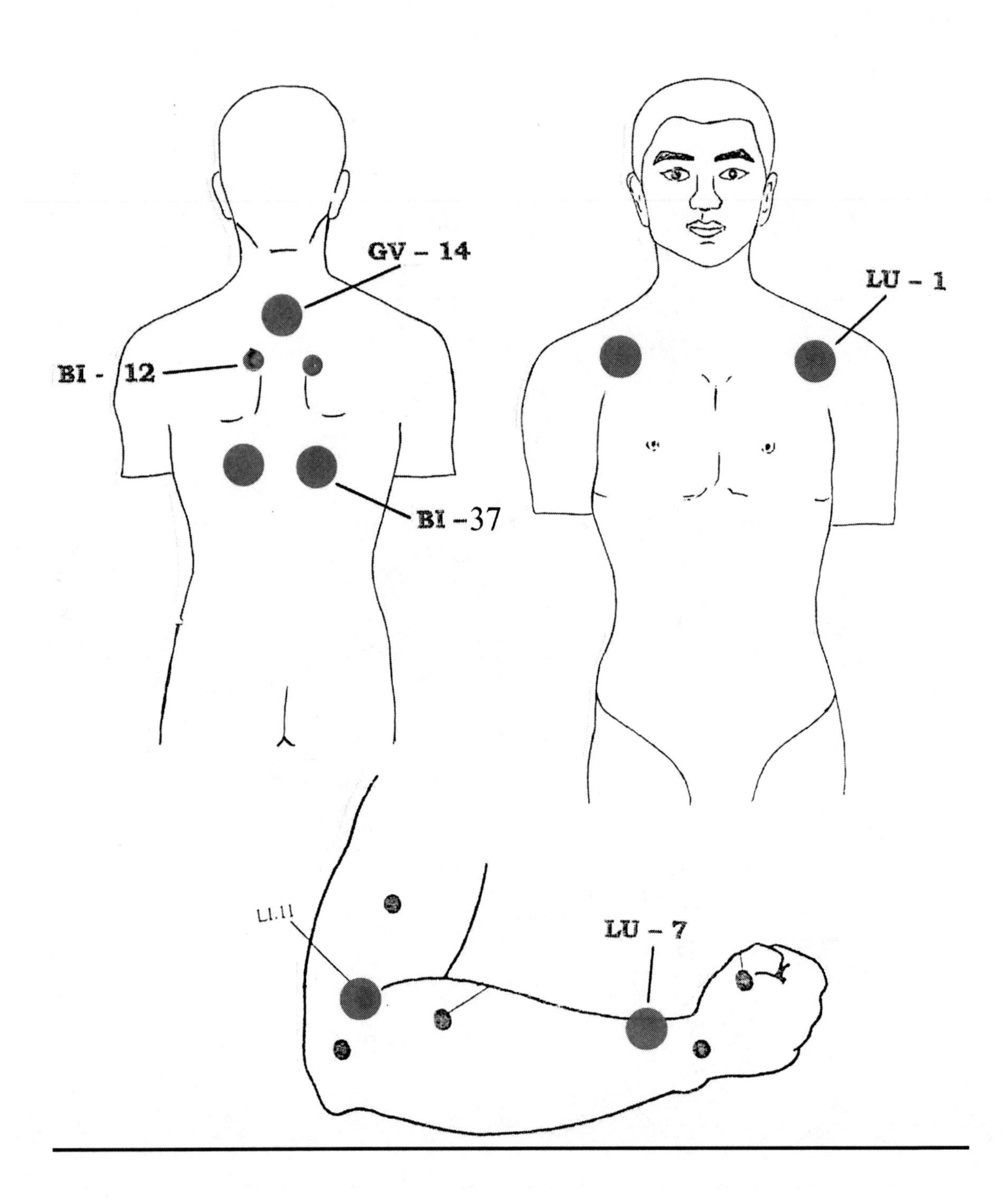

Respiratory System
Diagram 10c.
Common Cold and Flu

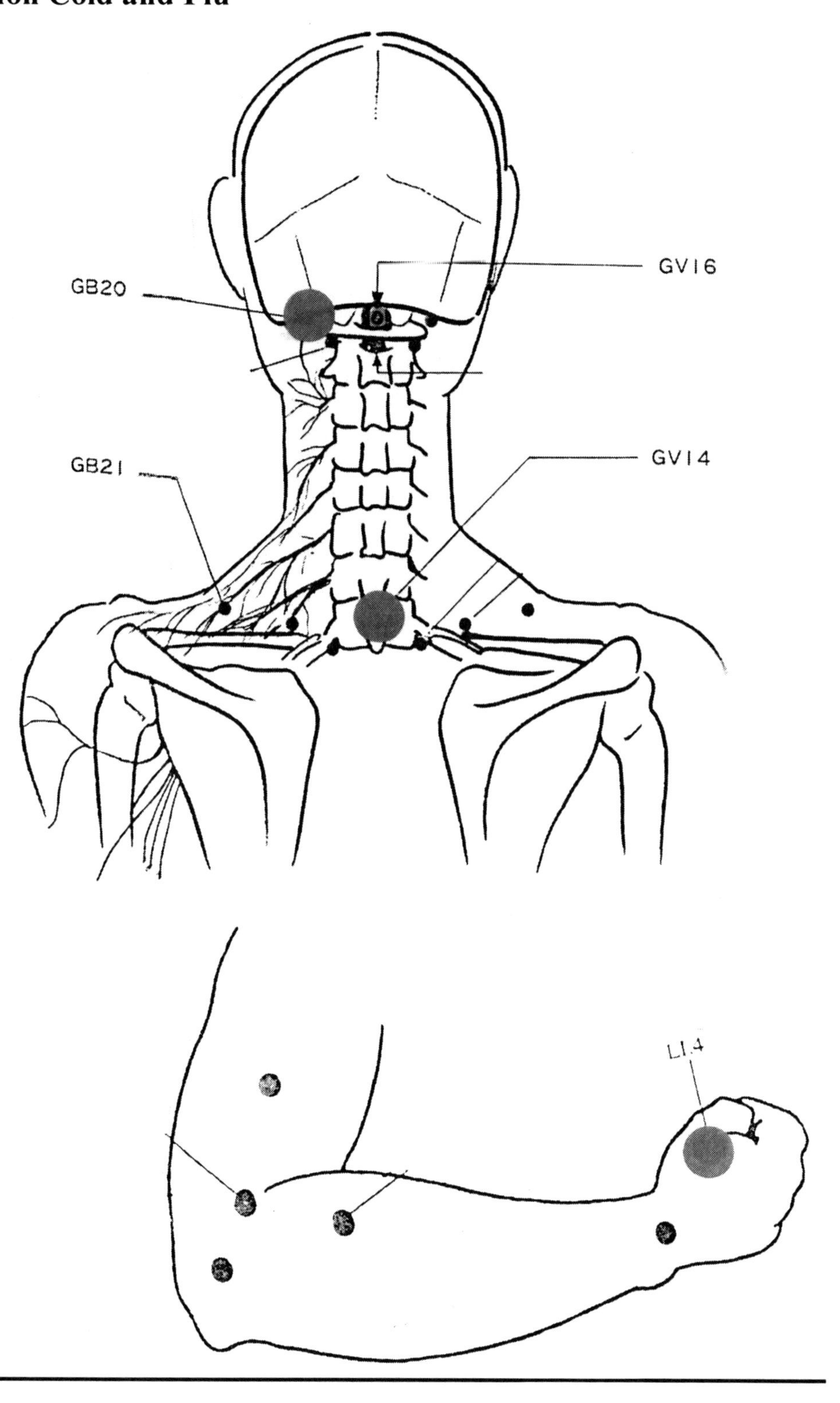

Skin
Diagram 12a.
Acne

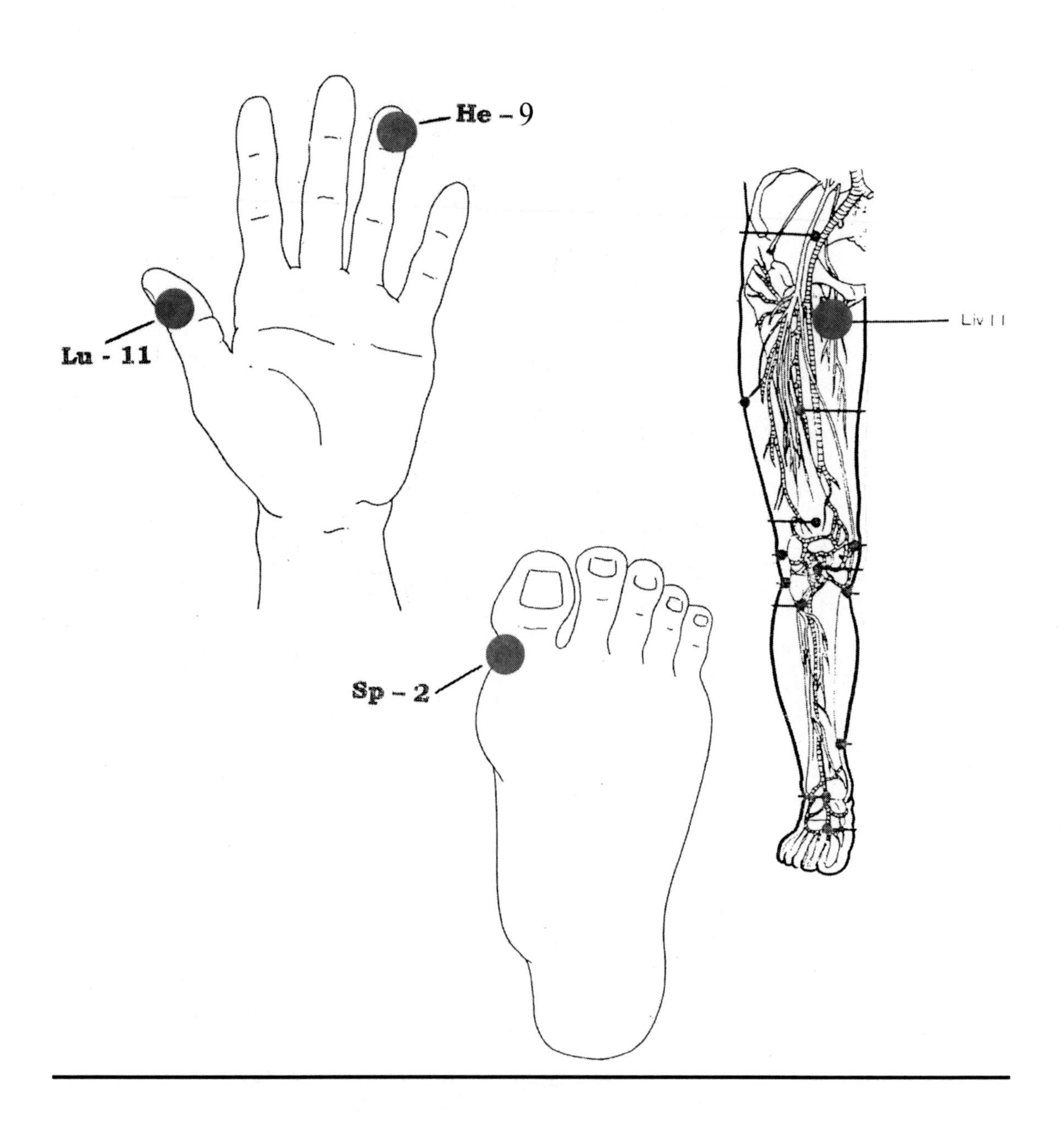

Skin
Diagram 12b.
Allergy: Hives, Eczema and Neurodermatitis

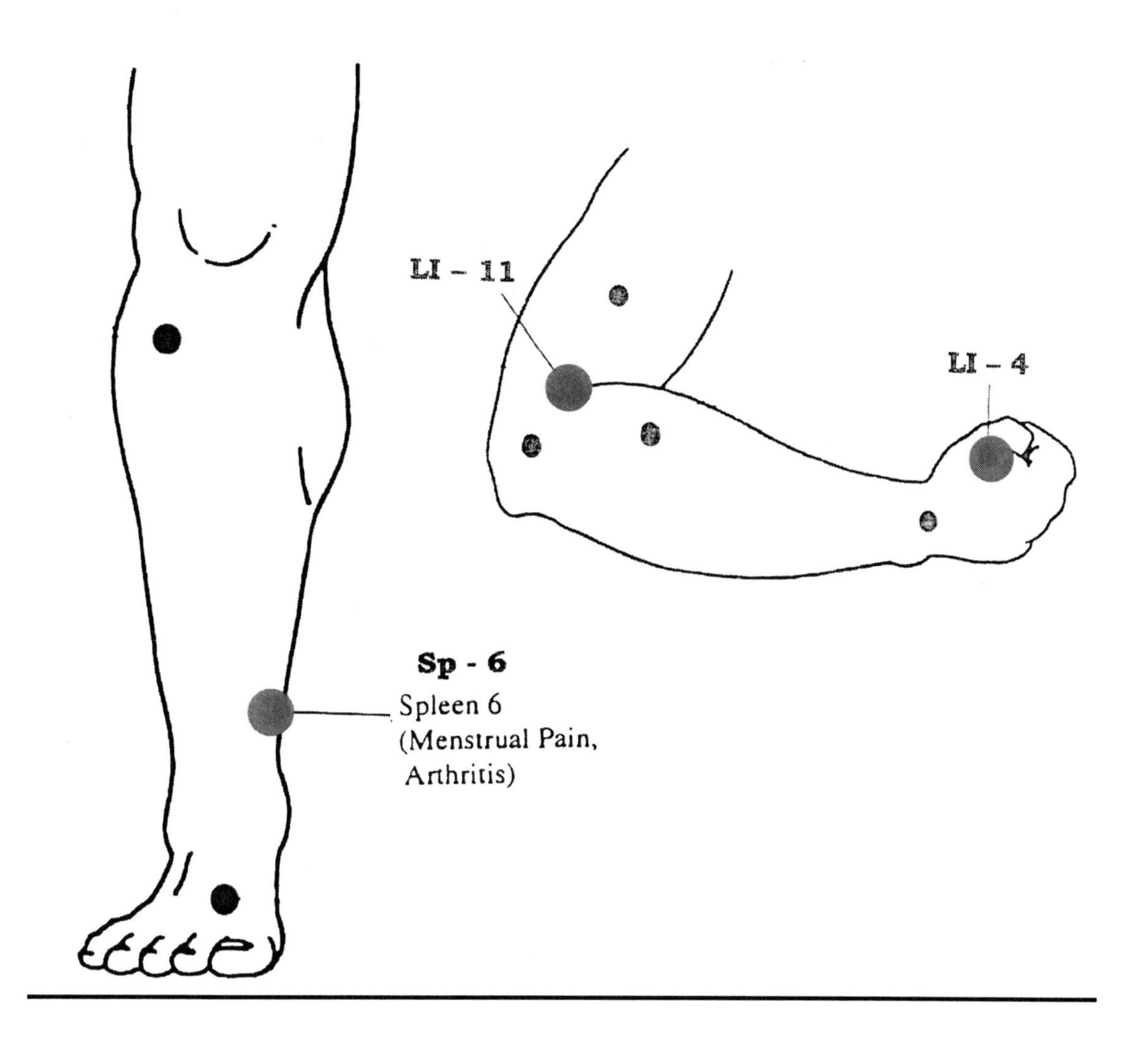

13. TENS for Facelifting (Muscle tightening and Weight Reduction)

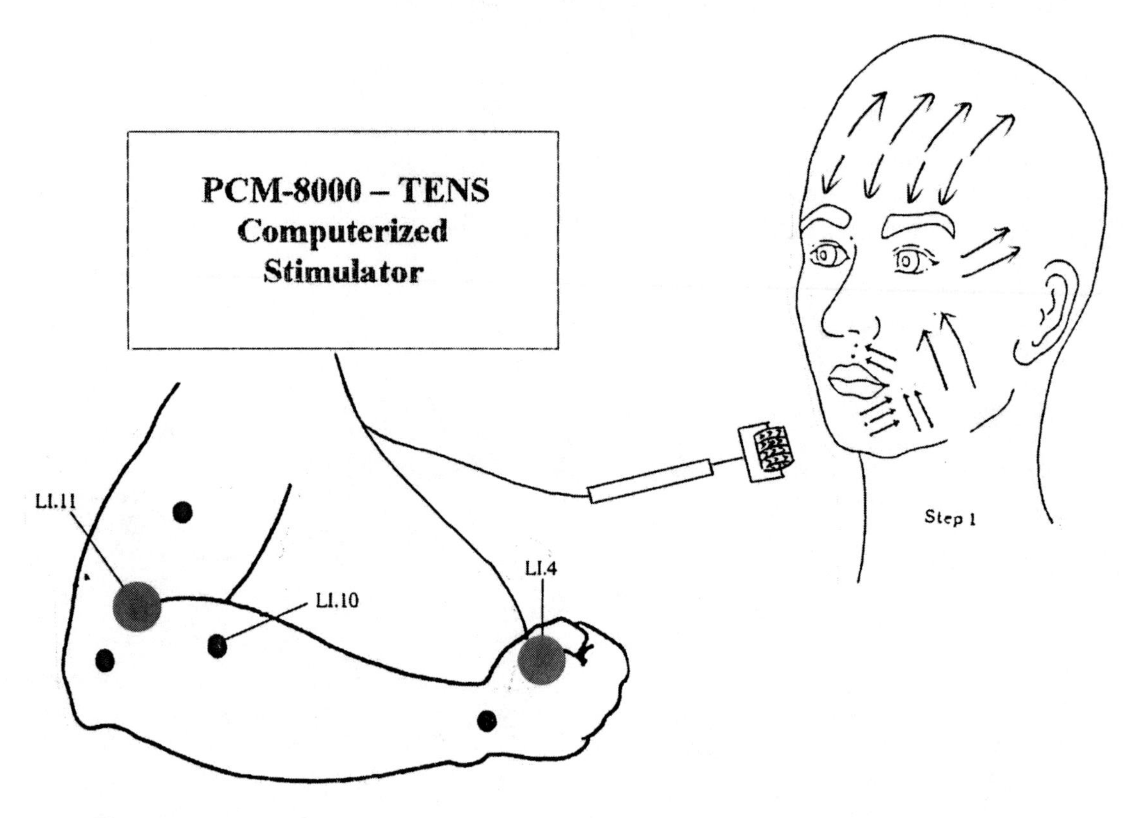

- **Use program – 8**
- **Intensity – comfortable level**
- **Slowly rolling 36 times to 100 times at each area on the face & neck (Total times – 60 minutes, use a skin cream e.g. Vitamin A, C, E cream)**
- **Do 2 to 3 times per week for 10 to 15 times or once per month for maintenance**

Muscular System Legend

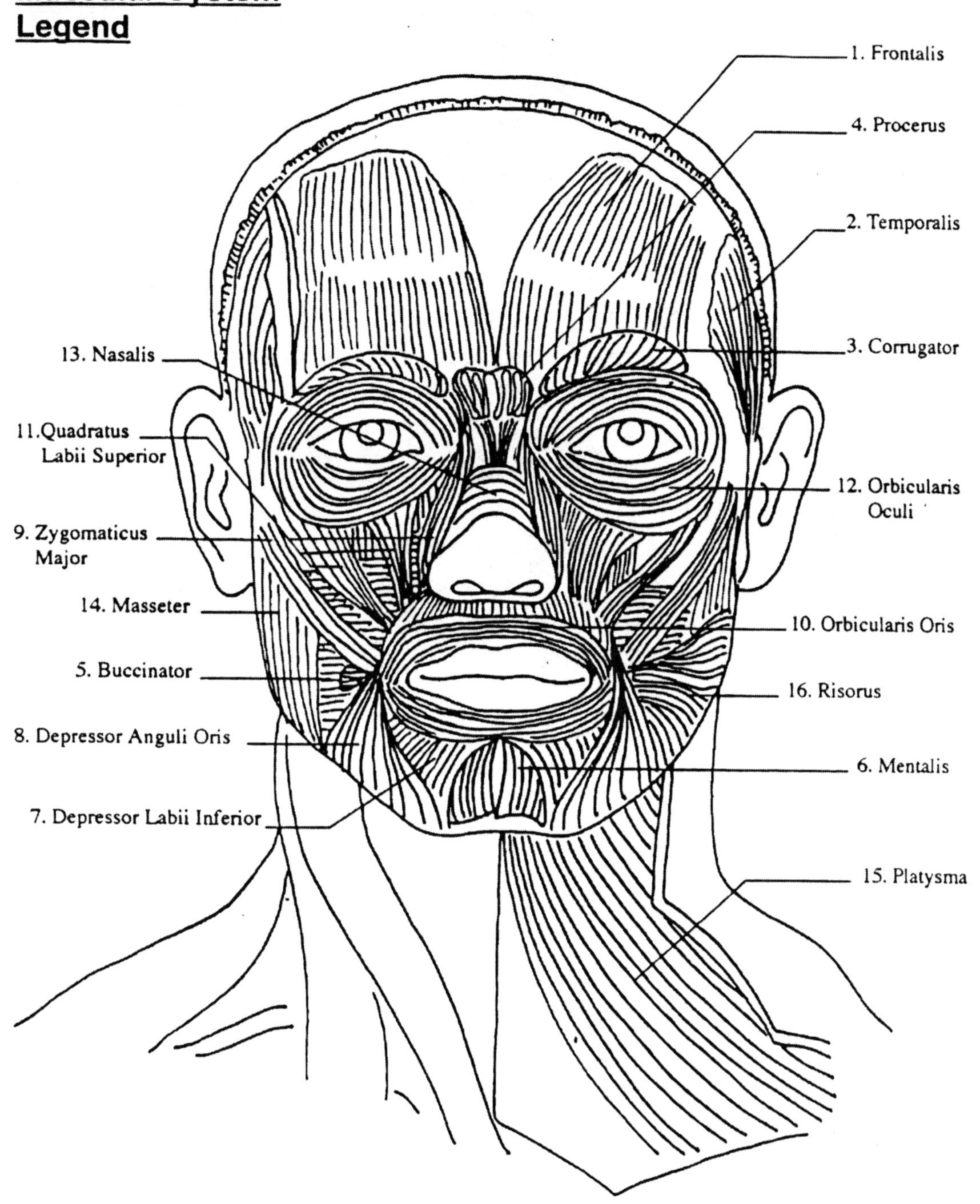

Medical Qi–Gong Meditation

By Dr. Richard Cheng, MD PhD

Introduction

I. Qi – Gong Meditation is a set of ancient Eastern training techniques that help promote physical and psychological well-being by teaching one to feel one's internal energy and harmonize it, achieving the "Qi", the vital energy of the body.

II. Meditation classes today are being filled by mainstream North Americans. It is offered in schools, hospitals, law firms, government and corporate offices, prisons and even at military school (West Point). Meditation is being recommended by more and more physicians as a way to prevent, slow or at least control the pain of chronic diseases like heart conditions, AIDS, cancer, infertility, depression, hyperactivity and attention-deficit disorder.

Scientific studies have shown that meditation can:

1. Boost the immune system (people who meditate have higher levels of immune cells known to combat tumors, e.g., breast and prostate cancers, etc.)
2. Lower blood pressure, slow the heart rate, promote heart health (by increase beat-to-beat variations) and enhance success in heart surgery.
3. Reduce chronic pain syndromes and stresses.
4. Help asthma, diabetic mellitus, and skin diseases such as eczema, psoriasis and other immune diseases.
5. Reduce degeneration or aging of the brain (those who meditate have bigger brain sizes seen in functional MRI).
6. Significantly activate the left prefrontal cortex brain cells in Qi-Gong practitioners (measured by spectrometer). This activation has been shown to be associated with increased levels of happiness, energy and optimism.
7. Qi-Gong can restore the human body back to healthy <u>Fractal</u> state

Techniques of Qi- Gong Meditation

The basic aim of Qi-Gong meditation is to silence the thinking mind and to shift the awareness from the rational to the intuitive mode of consciousness. Through the many forms of Qi-Gong meditation, the silencing of the rational mind is achieved by concentrating one's attention on a single item, like one's breathing, the sound of music or the visual image of a flower. Other schools focus the attention on body movement, which have to be performed spontaneously without the interference of any thought.

a. **Breathing Technique** – can be done sitting, standing or lying down. Paying attention to one's breathing: diaphragmatic breathing is to pay attention to the lower abdomen, bulging out on inhalation (diaphragm lowers) and sucking in on exhalation (diaphragm rises). By focusing the mind on the abdomen, one will feel a huge soothing heat or warmth expanding inside the abdomen. This warmth can be directed to flow from the base of the spine to the skull and back down to the abdomen. This is called the micro-cosmic orbit (refer to the diagram below). If one continues to direct this flow of energy, one will achieve a state of ultimate happiness and satisfaction.

b. **Dynamic Qi-Gong Meditation** – Simple body movements which are performed spontaneously without the interference of any thought. The rhythm of the movement can lead to the same feeling of peace and tranquility, while increasing circulation and strengthening the body part.

c. **Body Scan** – massage the acupuncture points; meridians or body parts, using your own hands or life energy, "Qi" or "feelings".

d. **Walking Meditation**: involves focusing on the sensations in the feet or legs, or alternatively, feeling the whole body moving. The experience of walking can integrate with breathing, or just feel the sensation of walking by making an effort to be fully aware as one's foot contacts the ground. This technique has been shown to reduce chronic pain syndromes, stress and boost the immune system, which can fight many kinds of cancer.

e. An instrumental DVD is available

Microcosmic Orbit

Internal Qi flow through Governing (GV) and Conceptive (CV) Vessels and Seven Chakras (1-7)

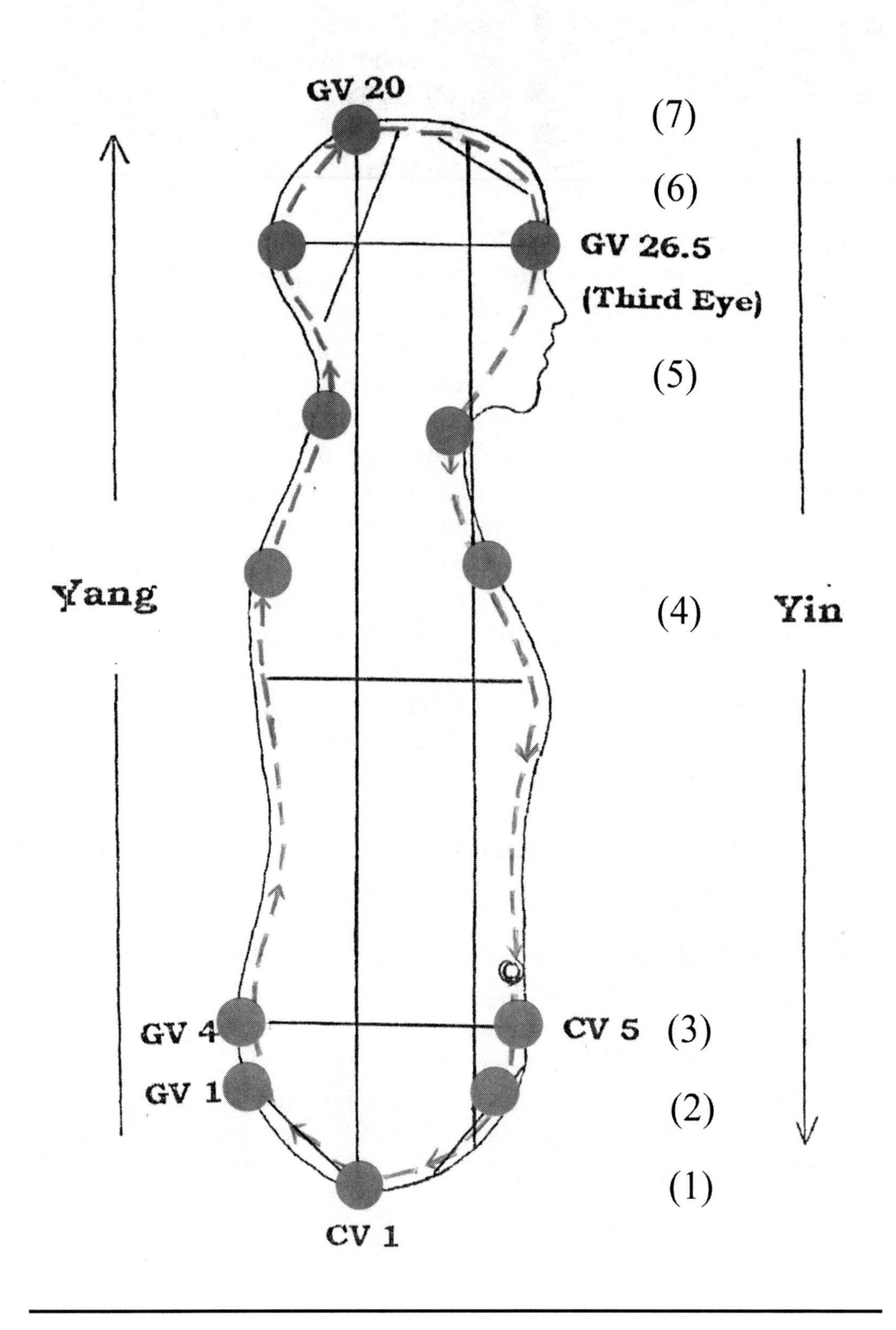

Appendix I

Diagram 1.

Lung Meridian

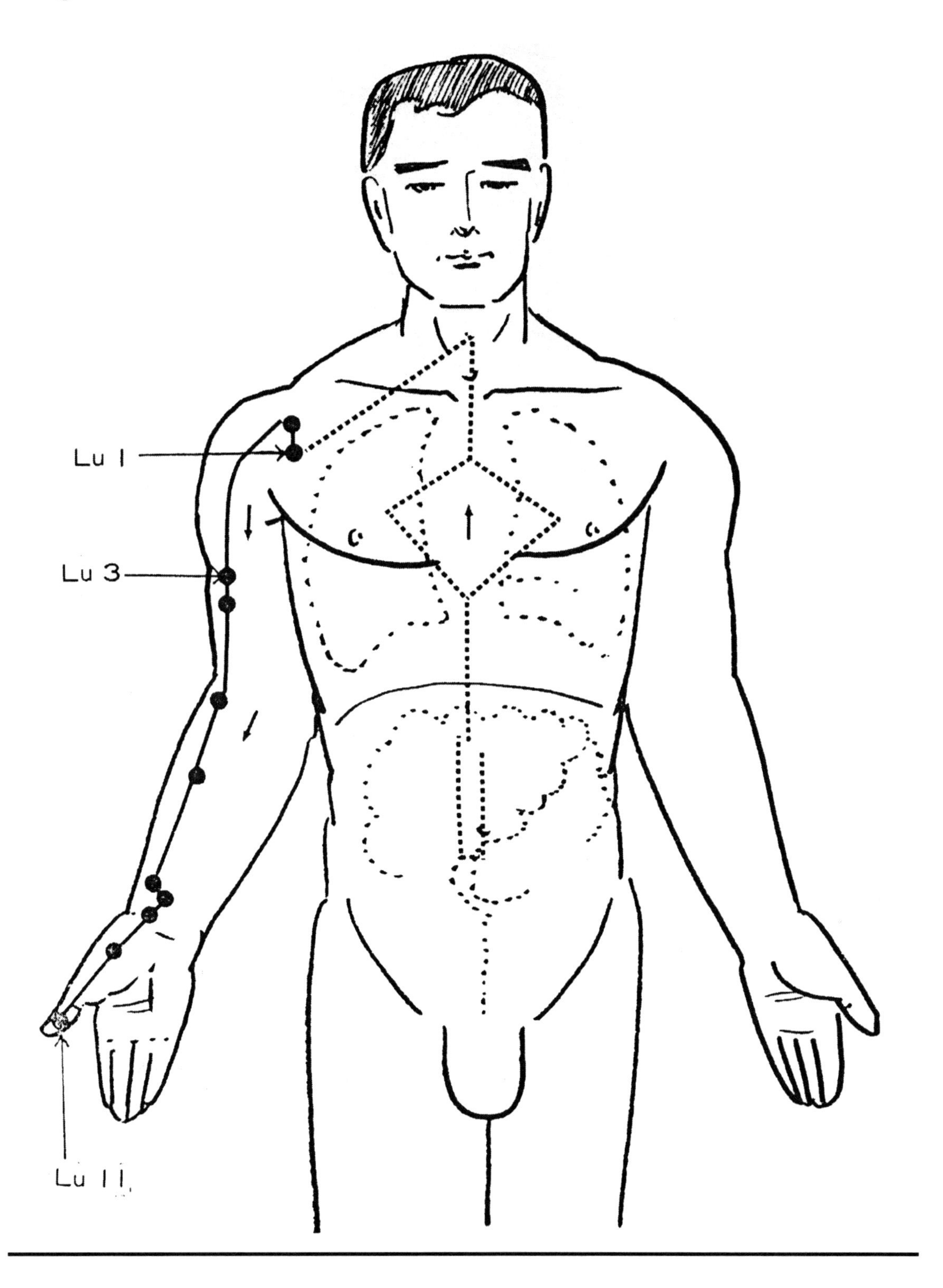

Appendix I
Diagram 2.
Large Intestine Meridian

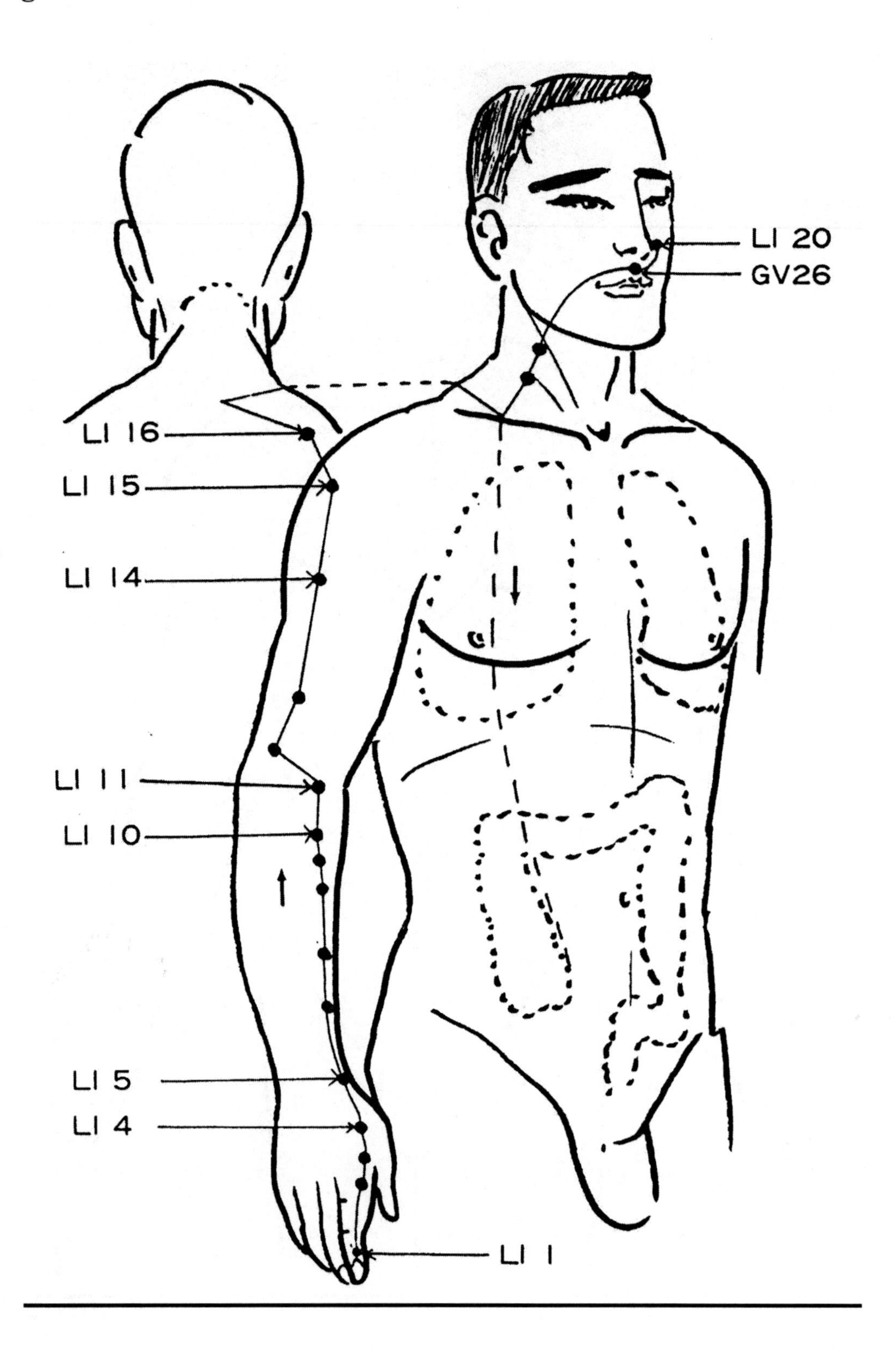

Appendix I
Diagram 3.
Stomach Meridian

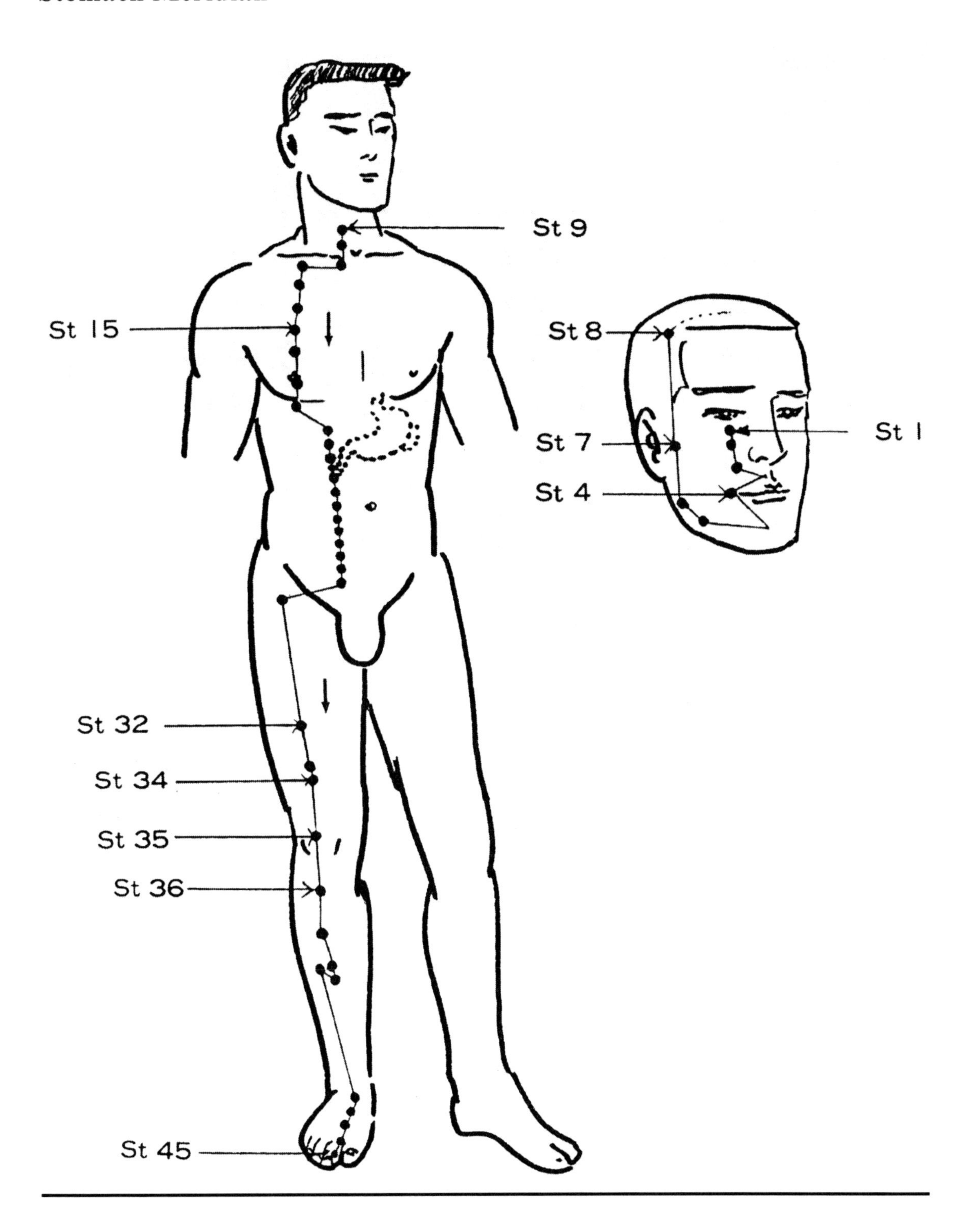

Appendix I
Diagram 4.
Spleen Meridian

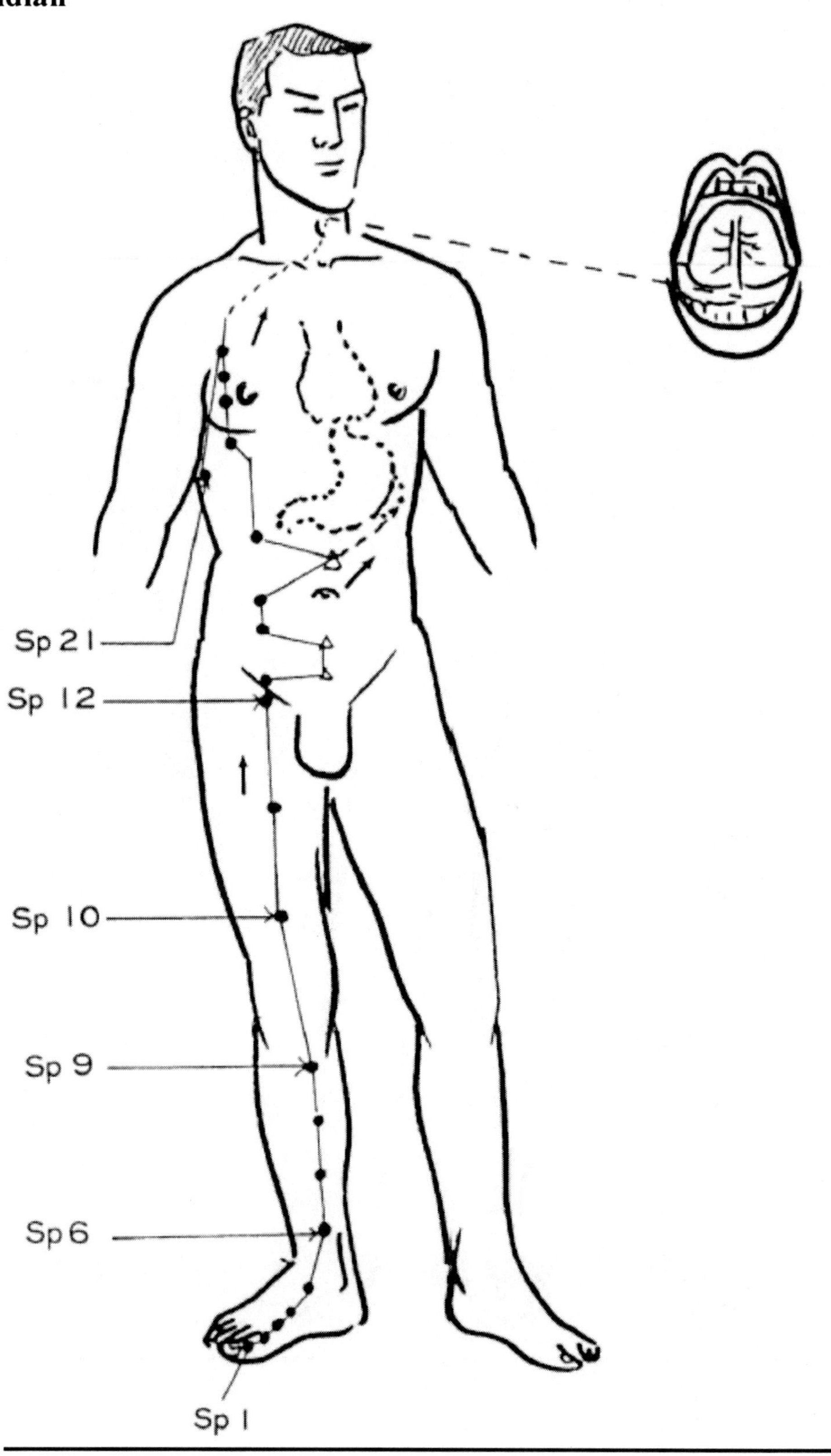

Appendix I
Diagram 5.
Heart Meridian

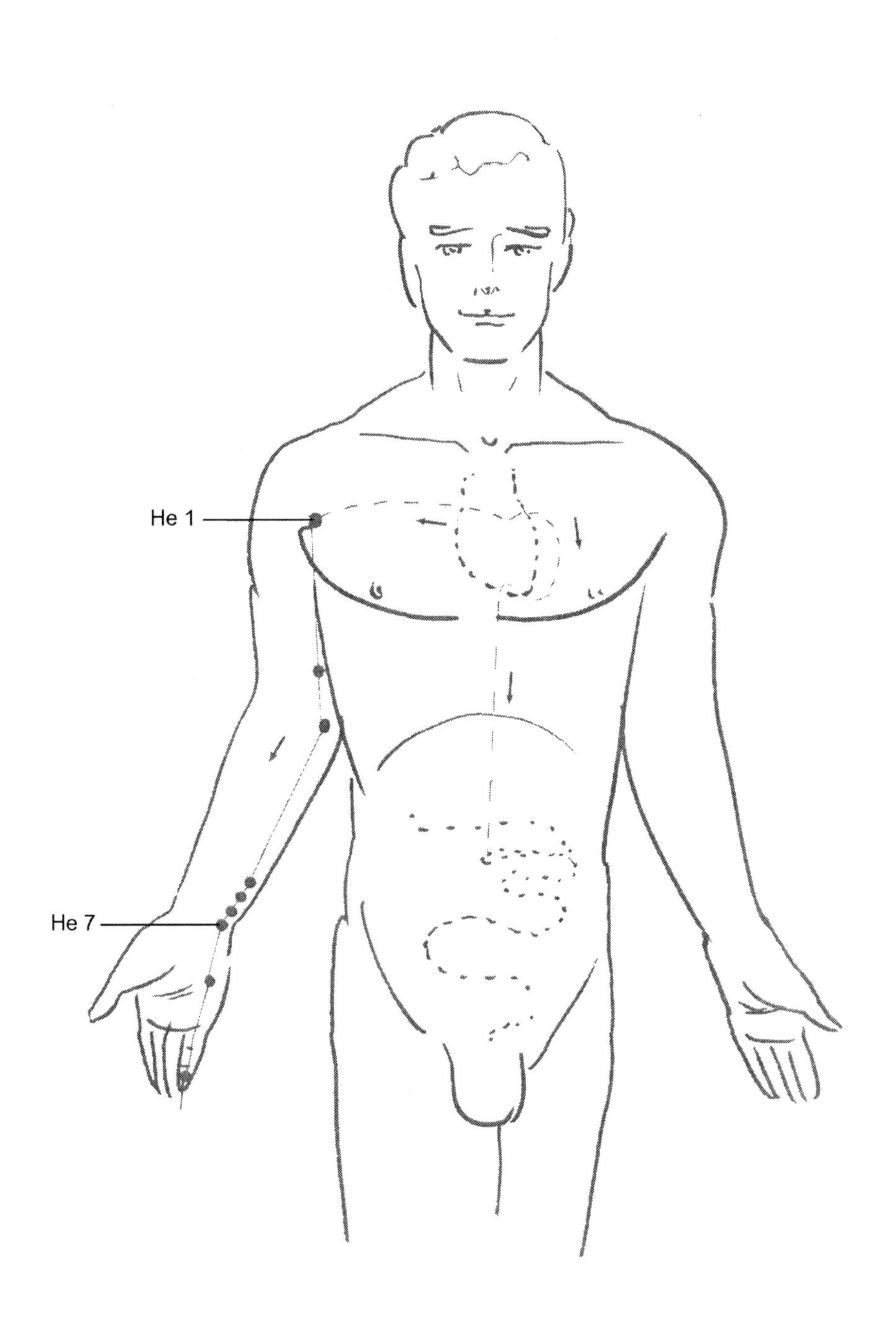

Appendix I
Diagram 6.
Small Intestine Meridian

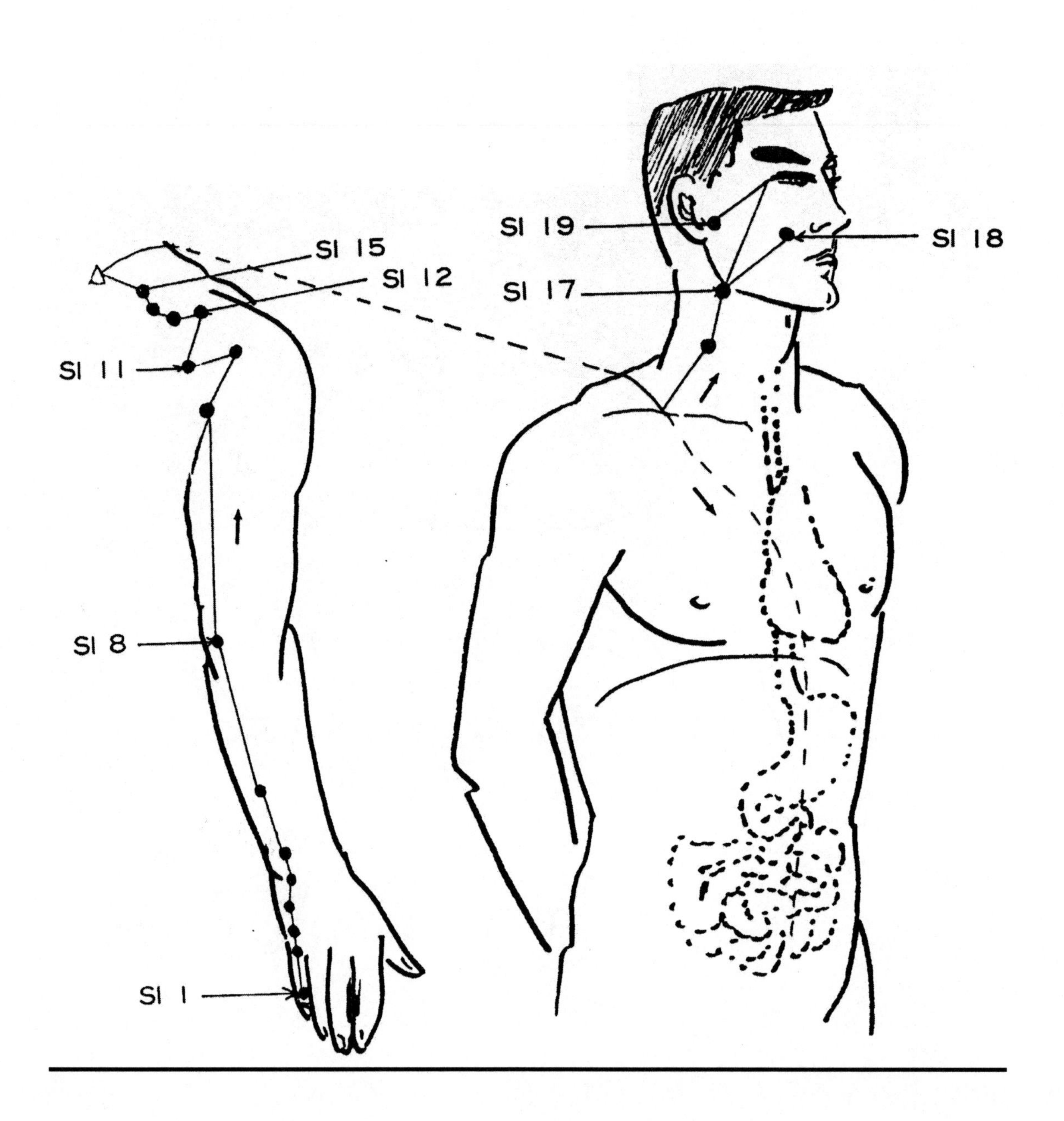

Appendix I
Diagram 7.
Urinary Bladder Meridian

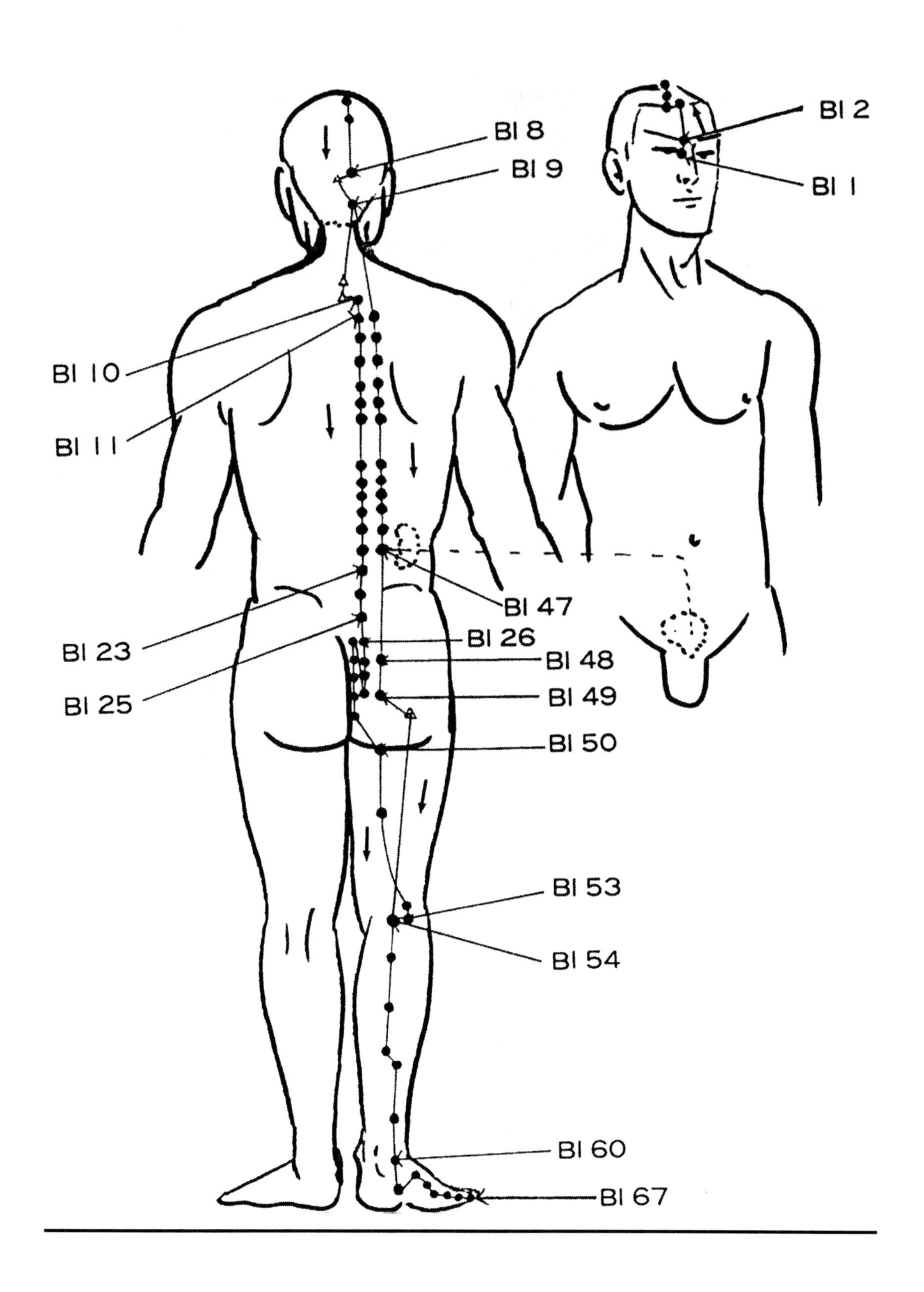

Appendix I
Diagram 8.
Kidney Meridian

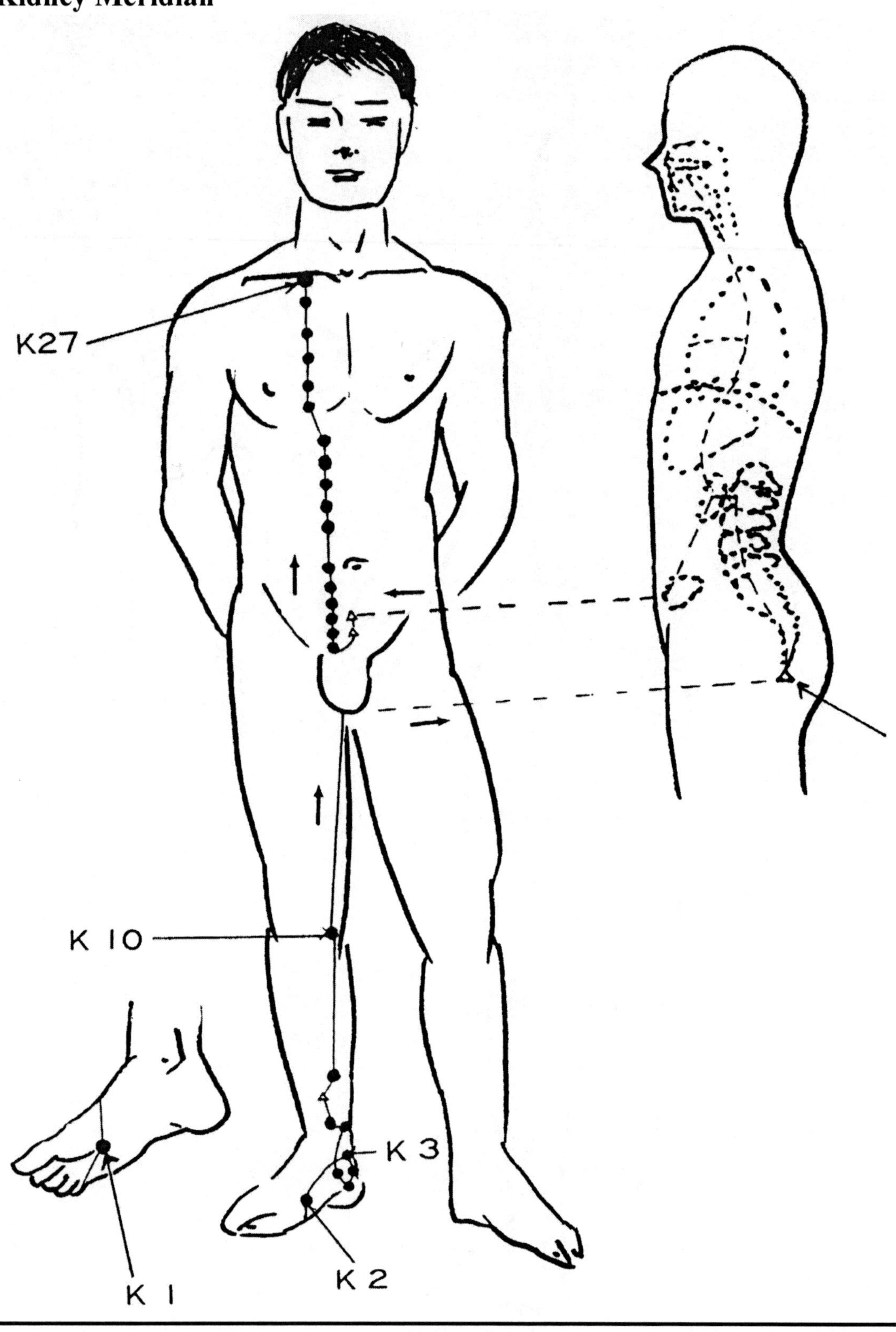

Appendix I
Diagram 9.
Pericardium Meridian

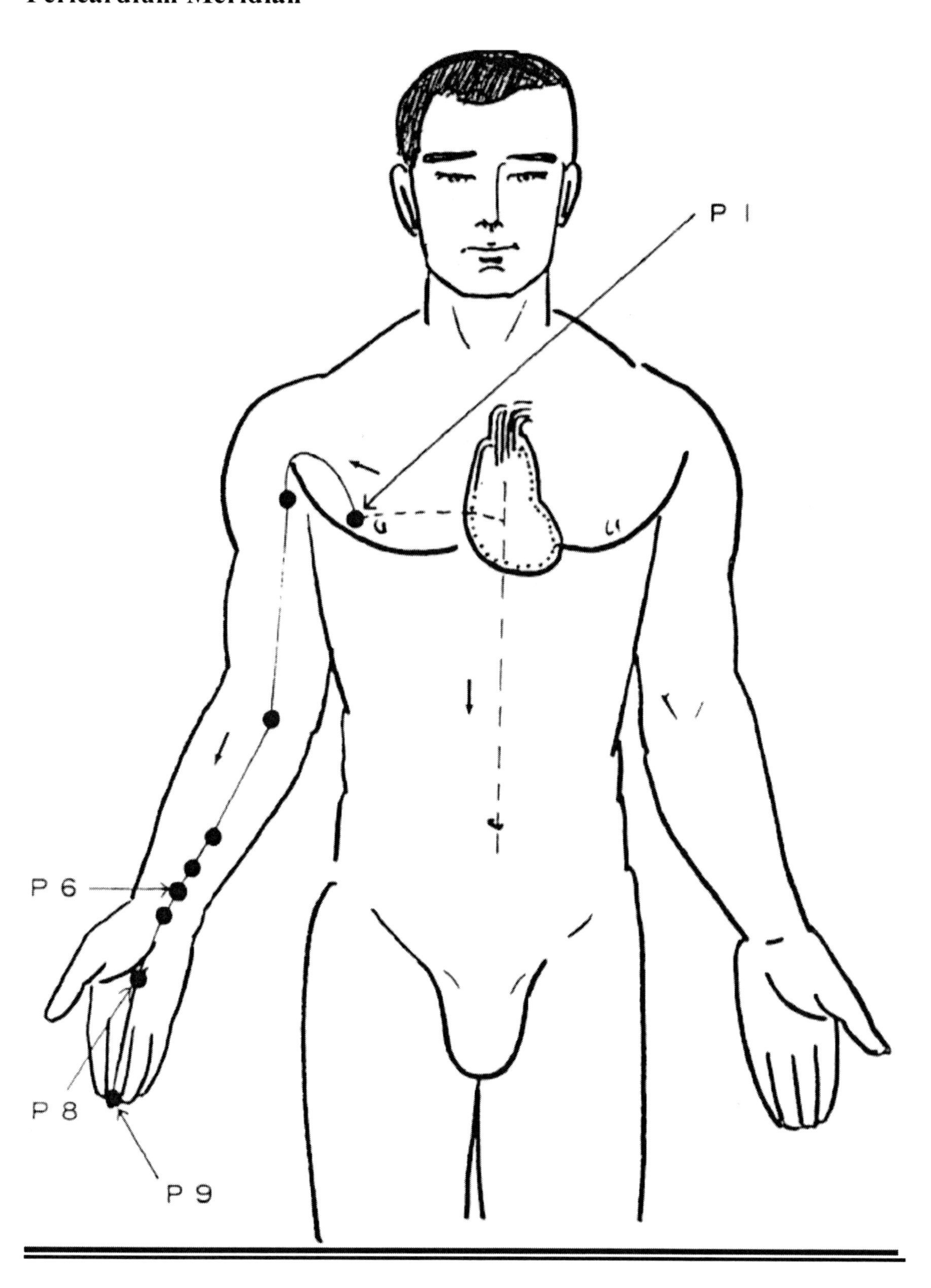

Appendix I
Diagram 10.
Triple Warmer Meridian

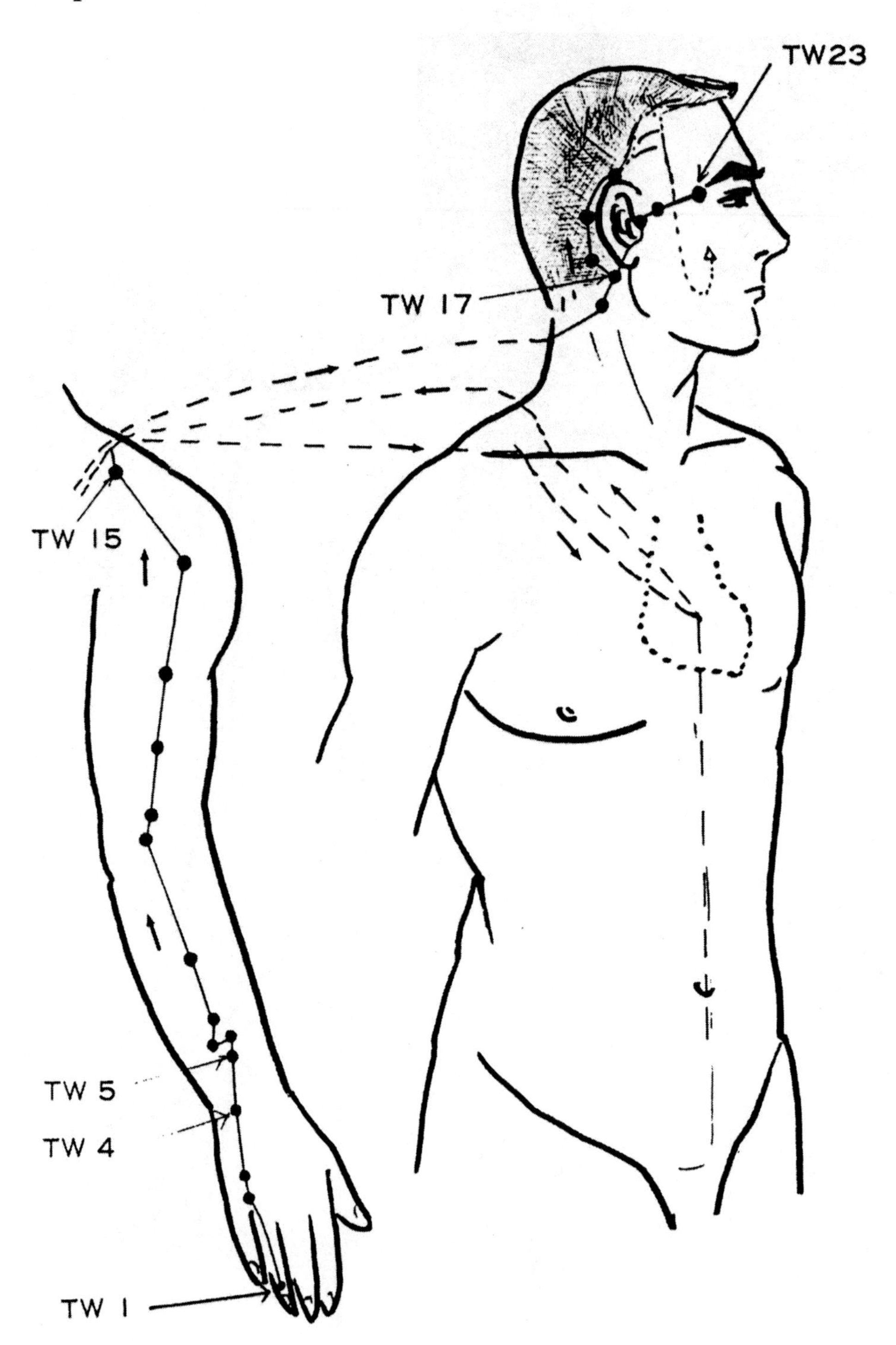

Appendix I
Diagram 11.
Gall Bladder Meridian

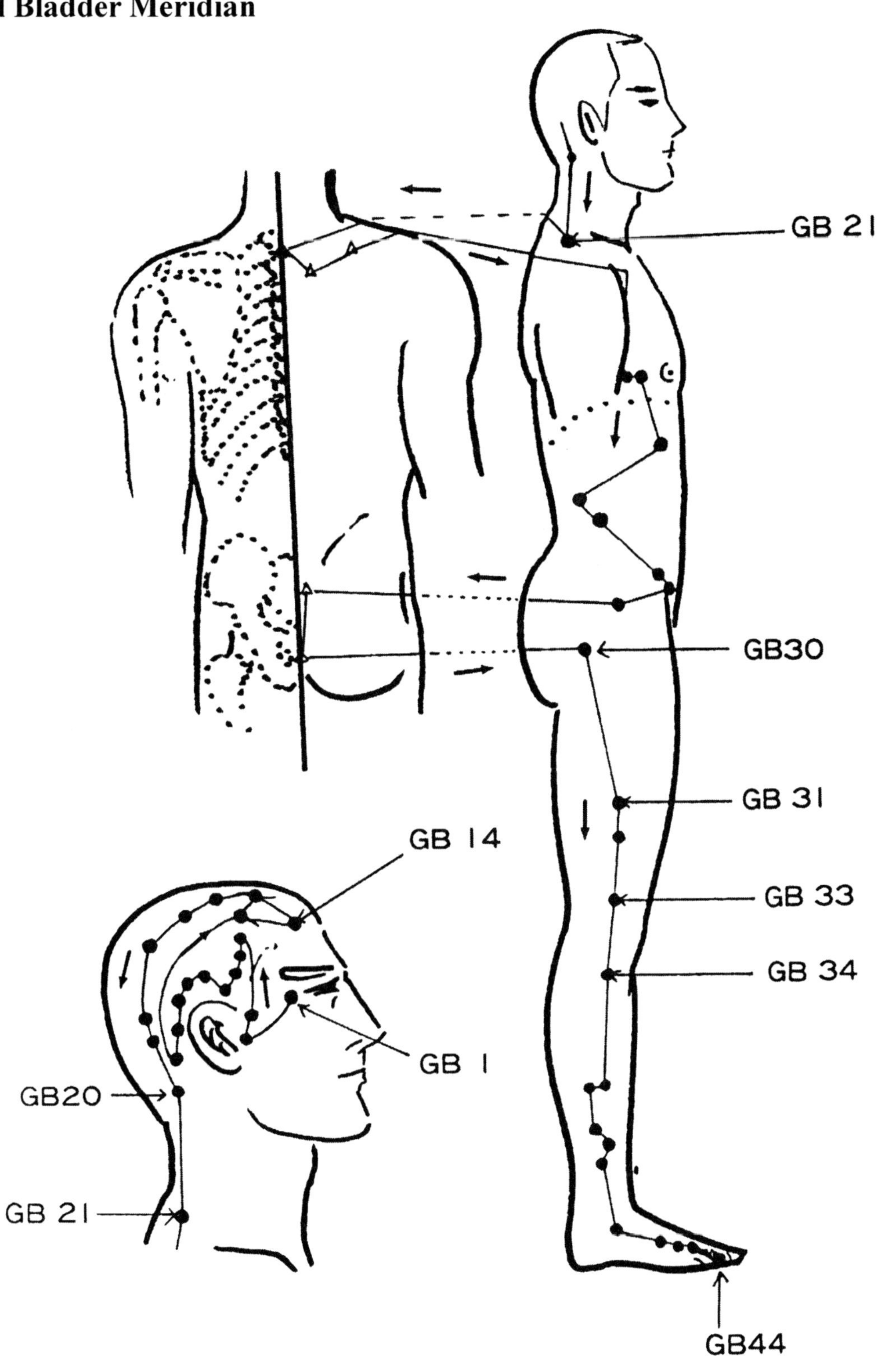

Appendix I
Diagram 12.
Liver Meridian

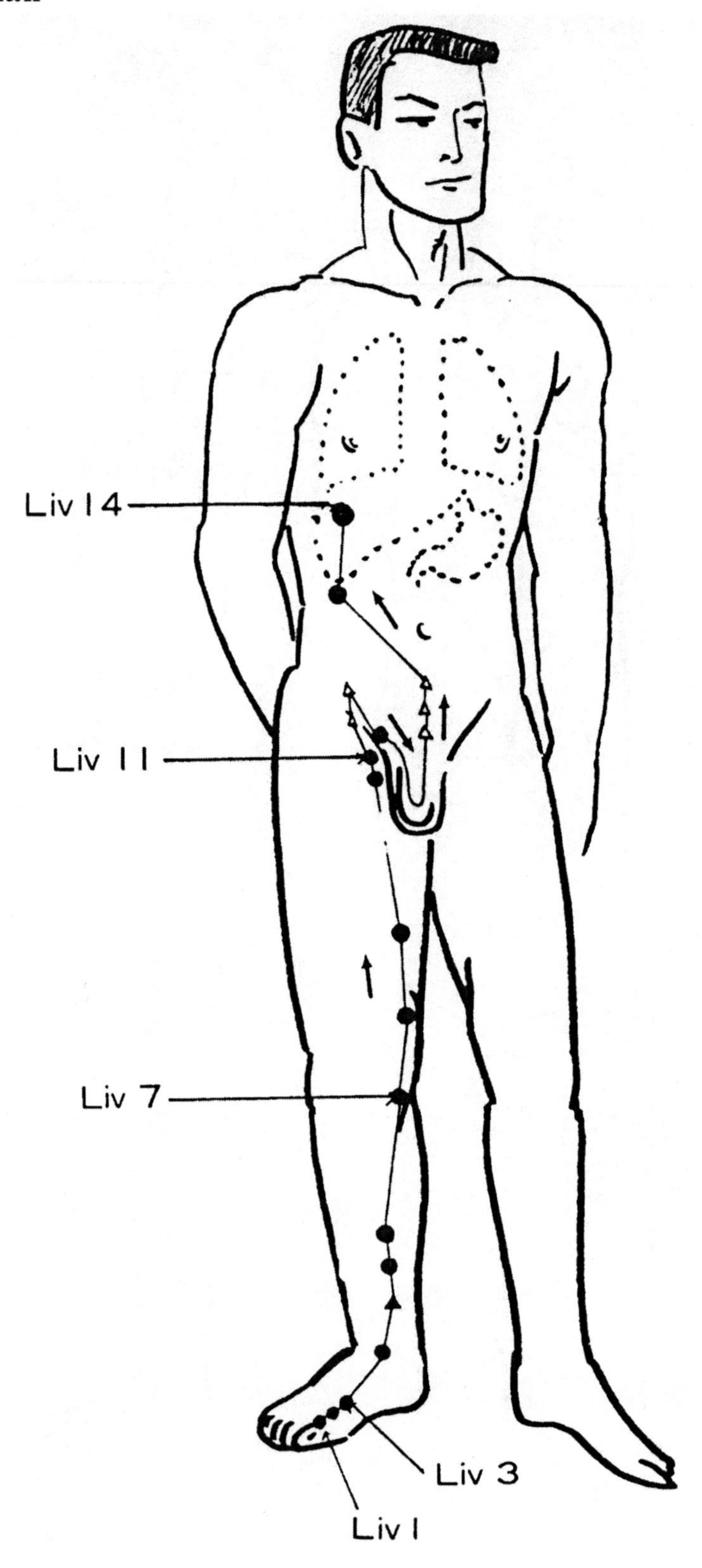

Appendix I
Diagram 13.
Governing Vessel (Meridian)

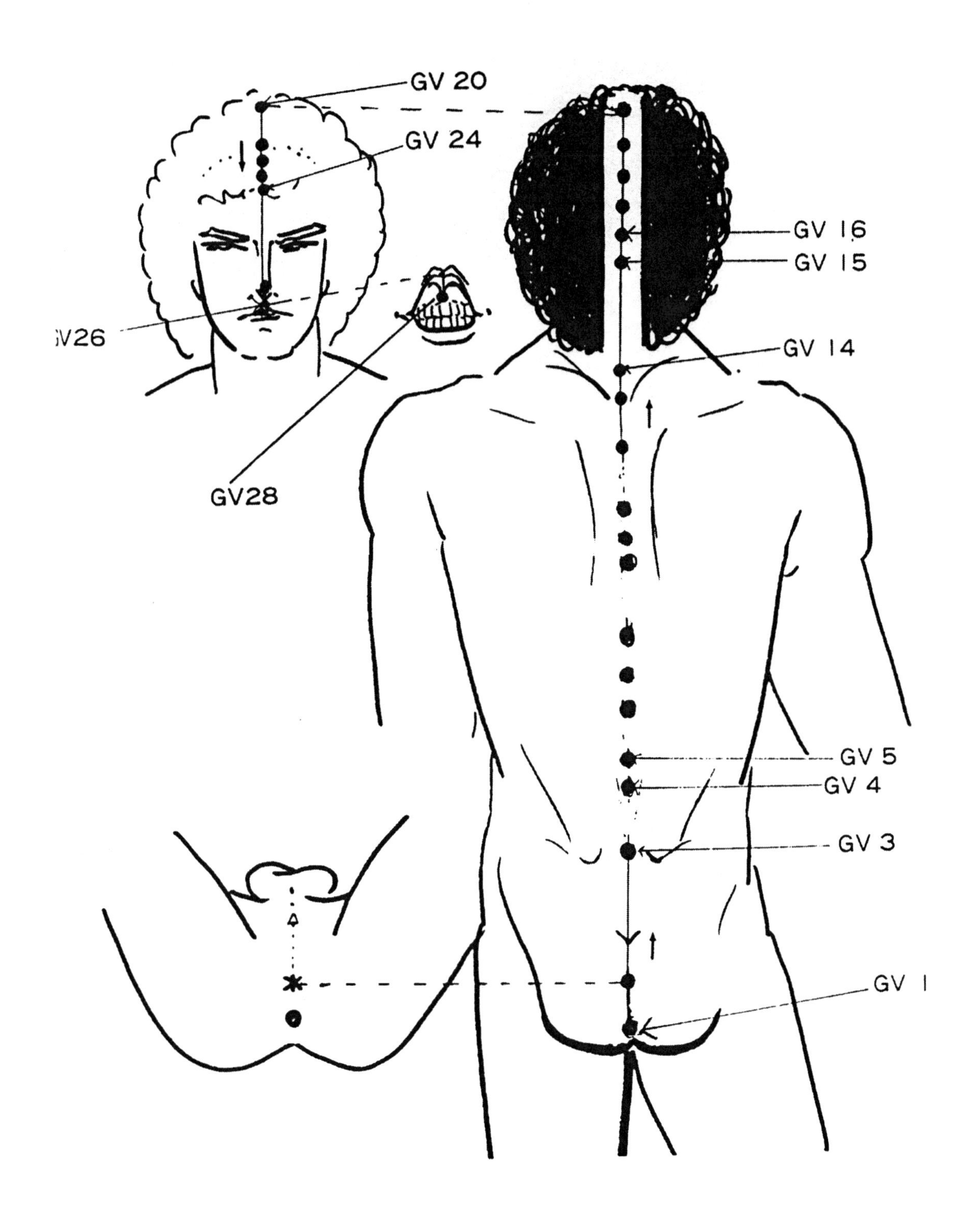

Appendix I
Diagram 14.
Conceptive Vessel (Meridian)

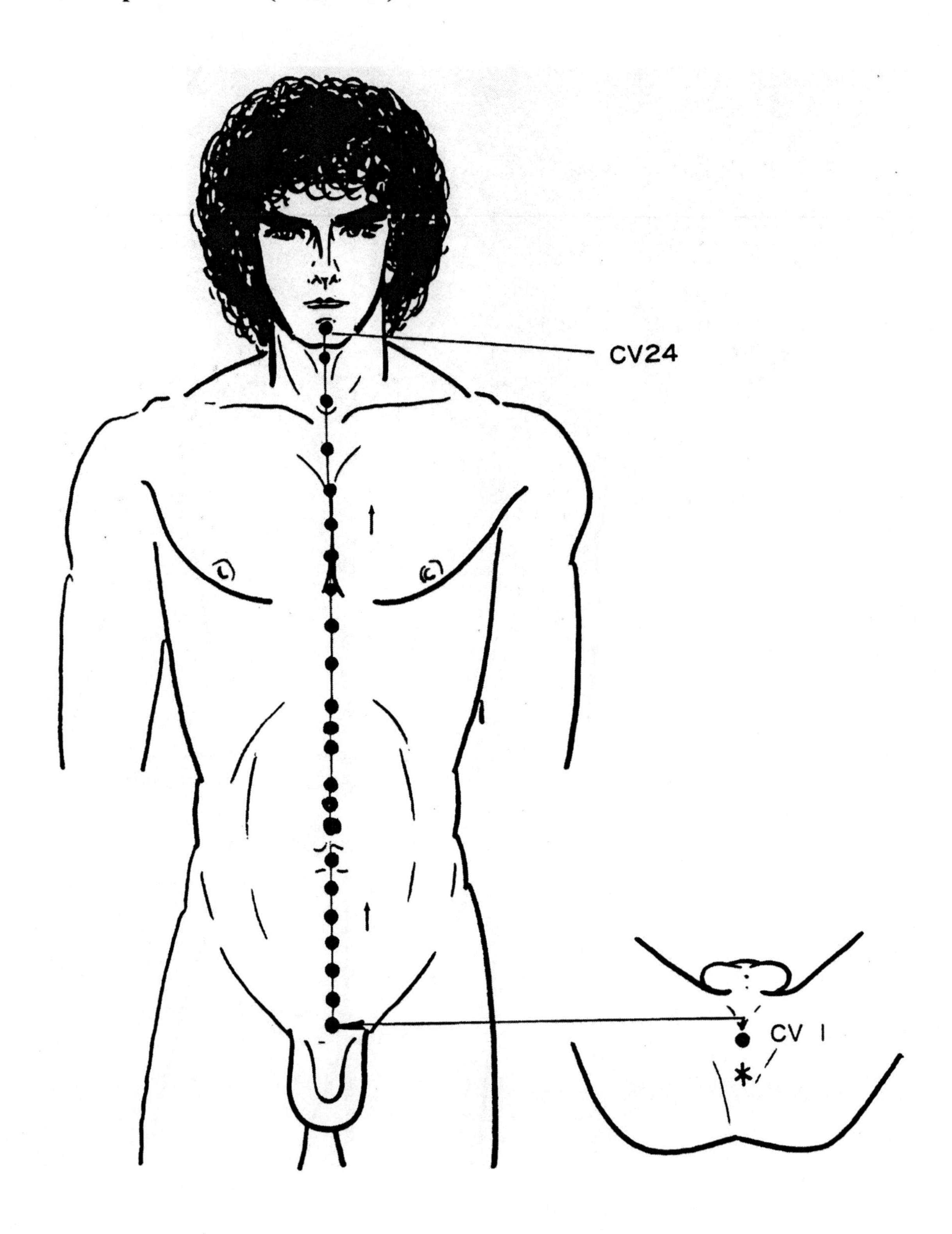

Appendix I
Diagram 15.
Ear Points

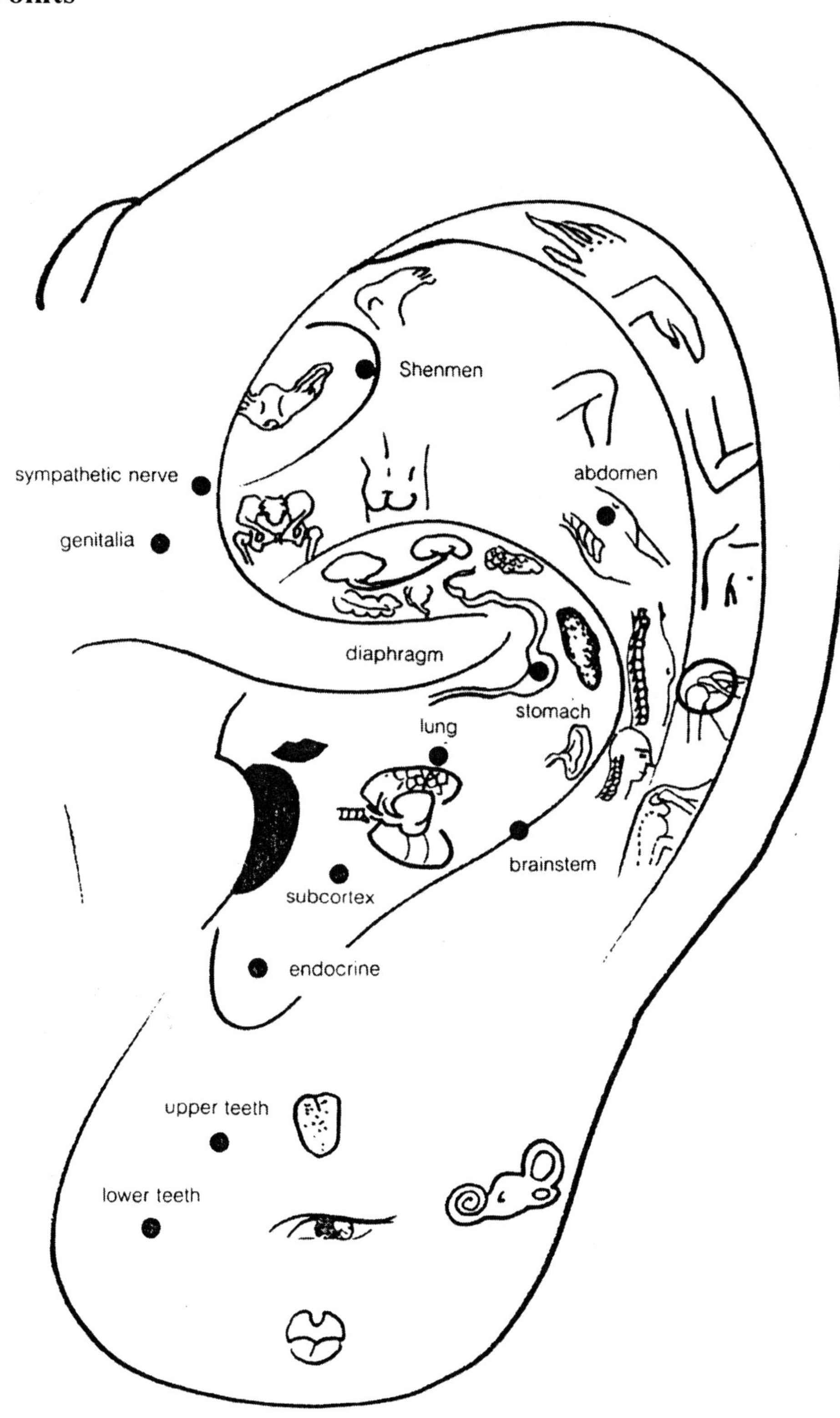

Appendix II
Illnesses Treated by Acupuncture

Acupuncture has been used to treat various illnesses, and often used in conjunction with other techniques or treatments. For interest, most of the traditional uses of acupuncture are listed below:

Disorder	Acupuncture Points	Ear Points
Abdominal pain	St–36, 10,12,P–6	Stomach, small intestine, sympathetic
Abortion	Ll– 4, Sp–6, CV–2, 3, 4, BI–60	
Acne	Sp–2, H–9, Ku–11, Liv–11	
Addiction	LI–4, H–7, St–36, P–6	Shenman, lung
Alcoholism	H–7, St–36	
Alopecia	BI–54, GV–20, 24	Intestinal secretion, lung, occiput, kidney
Amenorrhea	H–7, LI–4, Sp–6, GV–2,3,4, CV–4	Shenman, uterus, ovary, internal secretion
Amnesia	H–7 , P–6, K–1, St–36	Amnesia point
Angina pectoris	P–6, H–7, St–36, BI–15	
Anorexia	St–36, H–7, CV–9, 12	Stomach, small intestine
Anuria	CV–1, 6, K–1, BI–23, St–36, GV–2,3,4	
Anxiety	H-7 , P–7, LI–4	Shenman, heart, stomach
Apnoea	K–27, LI–4, GB–40	
Apoplexy	CV–24, GV–20, 26, St–36, GB–14, 20	Shenman, adrenal gland
Appendicitis	LI–11, St–44	Appendix, large intestine, shenman
Arteriosclerosis	St–36, GB–39, P–9	
Arthritis	Use points across the joints for local problems; General treatment: LI–11, St–36, LU–5, P–6, Sp–6	
Ascites	St–36, 45, CV– 6, 9, 12, Sp–6	Stomach, shenman, liver
Asthma	LU–5, 7, 9, LI–4, CV–16, 17	Shenman, lung, adrenal
Autoimmune disorders e.g., AIDS, psoriasis, rheumatoid arthritis, asthma, lupus, etc.	P–6, He–7, GV–14, LI–4, LI–11, St–36	
Bell's palsy	LI–4, St–4, 6	
Blenorrhagia	GB–1, 2, CV–1, Liv–1	

Disorder	Acupuncture Points	Ear Points
Blepharospasm	St–1, He–7	Eye
Blurred vision	SI–6, UB–2, GB–1, GB–14	
Bronchitis	St–36, LI–4, Lu–5	Lung, asthma, shenman
Cardiac arrhythmia	P–6, H–7	Shanman, heart, sympathetic
Carpal tunnel syndrome	H–7, LI–11 Electrical stimulation for 30 min	
Cataract	LI–4, St–36, GB–20	
Cerebral hemorrhage	St–6, LI–4, GV–26, GB–20	Shenman
Cerebral ischemia	St–36, 45, GV–20, GB–44	
Cholecystitis	GB–34, CV–12, P–6	
Colitis	St–36, CV–6, 9, 12	Large & small intestine
Common cold	LI–4, GV–14, 16, GB–20	
Conjunctivitis	GB–20, GV–22, BI–1	
Constipation	St–25, 36, CV–1, Liv–1	
Contraception	Sp–6, LI–4	
Cough	Lu–5, 7, LI–4	
Cystitis	B–3, CV–4, Sp–10	
Deafness	LI–4, GB–15, T–17	
Diabetes mellitus	P–6, BI–60, St–33, Lu–5	
Diarrhea	St–36, Sp–6, 14	Large & small intestine, shenman
Dermatitis	LI–11, P–6, GB–31	
Dementia	St–45, LI–45, T–10	
Depression	H–7, GV–14, P–6	
Dysentery	St–36, P–6, CV–4, Li–11	
Dysmenorrhea	Sp–6, CV–1, 4, 7, LI–4	
Dyspnea	LI–4, Lu–5, 7, 10	Lung, adrenal gland
Dysuria	Sp–6, GB–34, CV– 4	
Edema	St–36, Sp–6, CV– 4	
Emotional	H–7, GV–20, P–6, K–1	
Endocarditis	Sp–4, P–6, K–2, GB–41	
Enterocolitis	CV–6, 10, GV–3, Sp–1	
Enuresis	Sp–6, CV–4, K–10	Urinary bladder
Epididymitis	CV–6, Sp–6, Liv–2	
Epilepsy	P–6, St–36, GV–14, BI–20	
Epistaxis (nosebleed)	LI–4, Lu–11, GV–14, 23	

Disorder	Acupuncture Points	Ear Points
Esophageal spasm	Sp–6, B–38, GB–20, LI–11, CV–22	
Facial pain	LI–4, St–1, 4, 7 (contralateral side).	
Fainting	GV–26, K–1	
Fatigue	St–36, BI–20	
Fever	H–9, GB–22	
Flatulence	St–36, GB–25	
Frigidity	CV–3, K–12, B–23, Sp–6	
Gastritis	St–36, P–6, BI–21, CV–12, Liv–13	
Goitre	LI–4, GB–21, T–10, P–6	
Heart attack	H–7, GV–26, P–6	
Heat stroke	LI–4, GV–14, 20, BI–40	
Hiccoughs	BI–20, CV–12, S –36	
Hives	LI–4, 11, Sp–6	
Hepatitis	Liv–3, St–36, BI–10, LI–11	Liver, shenman
Hypertension	St–36, GB–20, LI–11, Liv–3, H–7	Shenman, antihypertension
Hysteria	H–7, P–6, LI–4, Sp–6	
Hematemesis	B–17, Sp–1, T–5	
Hemoptysis	B–17, Sp–4, CV–12	
Hemorrhage	Cerebral Gv-20, H-7, LI-4, St-36 Ocular: P–6 Intestinal: GV–1, B–18, 35, CV–4 Stomach: CV-1, GV-1, CV-12, B-27 Postpartum: H-7	
Hemorrhoids	CV–1, GV–1, Sp–10, B–57, GV–20	
Hypersalivation	LI–18, B–41, K–18, GB–23	
Hypotension	B–15, Sp–6	
Ileitis	St–36, B–21	
Ileus	St–36, B–21, P–6, CV–22	
Impotence	Sp–6, CV–3, 11, B–23	
Insomnia	H–7, Sp–6, LV–10, insomnia point	
Intercostal neuralagia	P–6, points on local dermatome	
Intestinal obstruction	St.–36, CV–6, 14, Sp–6, B–21	
Jaundice	LI–11, St–36, GB–21, Liv–8, CV-12	
Laryngitis	LI–4, LU–6, GB–12, K–7	
Lethargy	L.I–11, St–36, GB–34	
Malaria	P–9, LI–11, GV–14, T–5, SI–3	

Disorder	Acupuncture Points	Ear Points
Meniere's disease	GB–20, LI–4, P–6, St–36, SI–19	
Menopause	Sp–6, LI–4, CV–4	
Menorrhagia	Sp–6, LI–4, CV–4	
Migraine	LI–4, GV–24, GB–14, 20, H–7	
Myocarditis	P–6, 9, T–6	
Nausea & Vomiting	St–36, P–6, CV–12	
Nephritis	LV–14, T–9, B–23, K–13, Liv–14, St–28, CV–13	
Neuralgia	Local acupuncture points	
Neurasthenia	GB–34, LI–11, St–36, H–7	
Neuropathy	St–40, CV–15	
Nocturnal emission	B–13, H–7, Liv–3, K–1, CV–36	
Obesity	St–36, SI–19	Stomach, lung
Osteoporosis	Sp–6, BI–57	
Otitis media	GB–41, St–36, LI–4, GB–20	
Parotitis	GB–20, LI–4, LU–7	
Pelvic inflammatory diseases	CV–4, Sp–6, St–36, Liv–5, Sp–10	
Pericarditis	P–6, B–16, CV–14	
Peripheral vascular disease	K.2, Sp–6, B–55, GB–34	
Peritonitis	Liv–14, Sp–1, LI–13, K–17, Liv–14, St–36, Sp–9	
Pharyngitis	LI–4, St–44, LU–7	
Pleuritis	K–9, GB–44, Liv–14, Sp-21, B–19, K–23, GB–36, CV–18	
Prolapse (rectal)	GV–1, CV–8	
Prostatitis	CV–3, K–7, Sp–10, GV–20	
Prostate Hypertrophy	K–3, CV–4, Sp–10, B–67	
Raynaud's disease	Sp–6, K–6, LI–11	
Renal colic (kidney stones)	B–23, Sp–6, K–3, B–47	
Restlessness	H–7, 9, P–6	
Rhinitis	LI–4, 11, 20, GV–14, 24, GB–20	
Schizophrenia	LI–4, GV–26, H–7	
Sciatica	GB–30, B–50	Shenman, sciatic nerve
Shock	P–6, GV–24, K–1	
Sinusitis	LI–4, GV–23, T–33	

Disorder	Acupuncture Points	Ear Points
Speech impairment	G–15, CV–23	
Sterility	S.p–6, CV–3, K–2, 12	
Stomach ulcer	St–36, P–6, CV–14	
Stomatitis	Center of the ulcer or the surrounding area	
Stroke	LI–4, GV–29, Sp–6, K–1	
Sweating (Hyperhidrosis)	CV-12, St–36, LI–4, LU–11	
Syncope	GV–20, 26, Sp–6, H–7, P–6	
Tonsillitis	LU–11, LU–7, H–7, LI– 4	
Tinnitus	GB–20, K–1, T–17	Inner ear, kidney, shenman
Tracheitis	LI-4, P–3, GV–12	
Urethritis	CV–6, B–23, Liv–13	
Urinary incontinence	LU–7, GV–4	
Urinary retention	CV–4, 6, Sp–10, Kid–11, St–37, B–54	
Urinary tract infections	Sp–6, St–36, B–32, Liv–8	
Varicose veins	GV–4, 14, GB–21	
Vertigo	LI–4, GB–20, St–40	
Vitiligo	Sp–6, GV–14, LI–11, LI–4, LU–7	
Vomiting	P–6, St–36, CV–12	

References – Acupuncture

1. Alvares D and Fitzgerald M (1999). Building blocks of pain; the regulation of key molecules in spinal sensory neurones during development and following peripheral axotomy. Pain. Supplement 6 (1999); S 71-S-85.
2. Basbaum AI, Fields HL (1984). Endogenous pain control systems: brainstem spinal pathway and endorphin circuitry. Ann Rev Neurosci; 7: 309-38.
3. Besson JM and Oliveras JL (1973). Analgesia from electrical stimulation of the periaqueductal grey matter in the cat: behavioural observations and inhibitory effects on spinal cord interneurons. Brain Res. 50: 441-6.
4. Cheng, RSS (1977). Electroacupuncture effect on cat spinal cord neurons and conscious mice: A new hypothesis is proposed. MSc Thesis, Zoology. University of Toronto.
5. Cheng, RSS (1981). Mechanism of electoacupuncture analgesia; an intricate system is proposed. PhD Thesis. University of Toronto.
6. Cheng, RSS and Pomeranz B (1979). Electroacupuncture analgesia could be mediated by at least two pain-relieving mechanisms; endorphin and non-endorphin systems. Life Sci. 25, 1957-1962.
7. Cheng, RSS and Pomeranz B (1980). Monoaminergic mechanisms of electroacupuncture analgesia. Brain Res. 215, no. 1-2, 77-93.
8. Cheng, RSS (1988). Neurophysiology of Electroacupuncture Analgesia, 119-136 in "Scientific Bases of Acupuncture" ed. by Bruce Pomeranz and Gabriel Stux, published by Springer-Verlag.
9. Cheng, RSS and Jussaume, R (1999). Pain Management and the mechanisms of Trancutaneous Electrical Nerve Stimulation (TENS) and acupuncture. Mature Medicine. Vo.2, No. 5:254-257.
10. Dubner R and Ren, K (1999). Endogenous mechanisms of sensory modulation. Pain. Supplement 6 (1999); S 45-53.
11. Han JS and Xie GX (1984). Dynorphin: Important mediator for electro-acupuncture analgesia in the spinal cord of the rabbit. Pain. 18:367-76.
12. Han JS (1988). Central Neurotransmitters and Acupuncture Analgesia, 7-34 in "Scientific Bases of Acupuncture" ed. by Bruce Pomeranz and Gabriel Stux, published by Springer-Verlag.
13. Jabbur SJ and Saade NE (1999). From electrical wiring to plastic neurons: evolving approaches to the study of pain. Pain. Supplement 6 (1999); S 87-S 92.
14. Mayer DJ, Price DD, Barber J et al (1976). Acupuncture analgesia: Evidence for activation of a pain inhibitory system as a mechanism of action in: Bonica JJ Albe-Fessard D, eds. Advances in Pain Research and Therapy. New York: Raven Press, 754-6.
15. Mayer DJ, Wolfle JL, Akil H et al (1971). Analgesia from electrical stimulation in the brainstem of the cat. Science. 174:1351-4.
16. Melzack R (1999). From the gate to the neuromatrix. Pain. Supplement 6, S121-126.
17. Petersen–Zeitz KR and Basbaum AI (1999). Second messengers, the substantia gelatinosa and injury–induced persistent pain. Pain. Supplement 6, S 5-12.
18. Pomeranz, B (1988). Acupuncture Research related to pain, drug addiction and nerve regeneration, 35-52 in "Scientific Bases of Acupuncture" ed. by Bruce Pomeranz and Gabriel Stux published by Springer-Verlag.

References - Qi Gong Meditation

1. Brazy JE. Cerebral oxygen monitoring with near infrared spectroscopy: clinical application to neonates. J Clin Monit 7:325–334, 1991.
2. 1st World Conference for Academic Exchange of Medical Qi-Gong. 1988. Beijing, PRC.
3. Cheng R. Text-Book of Medical Qi-Gong. pp 1–3, 2003 (in press).
4. Sancier KM, Hu BK. Medical Applications of Qi-Gong and Emitted Qi on Humans, Animals, Cell Cultures, and Plants: Review of Selected Scientific Research. Am J of Acupuncture. Vol.19, No.4:367–377, 1991.
5. Sancier KM. Medical applications of Qi-Gong. Altern. Ther. Health Med. 2:40–46, 1996.
6 Wallace RK, Physiological Effects of Transcendental Meditation, Science 167: 1751–1754, 1970.
7. Lee MS, Huh HJ, Jang HS, Han CS, Ryu H, Chung HT, Effects of Emitted Qi on *In Vitro* Natural Killer Cell Cytotoxic Activity. Am J of Chinese Medicine, Vol.29, No.1:17–22, 2001.
8. Fukushima M, Kataoka T, Hamada C, Matsumoto M, Evidence of Qi-Gong Energy and its Biological Effect on the Enhancement of the Phagocytic Activity of Human Polymorphonuclear Leukocytes. Am J of Chinese Medicine, Vol. 29, No.1, pp1–16, 2001.
9. Wu, WH and Bandilla E: Effects of Qi-Gong on Late–Stage Complex Regional Pain Syndrome. Alternative Therapies. 5:45–54, 1999.
10. Lazar, SW et al. Meditation experience is associated with increased cortical Thickness. Vol. 16, No.17:1893-1897, 2005.
11. Peng C-K, Mictus JE, Liu Y-H Khalsa, G, Douglas PS, Benson H and Goldberger AL, Exaggerated heart rate oscillations during two meditation Techniques. International Journal of Cardiology, 70 (1999) 101-107.
12. Davidson RJ et al. Alterations in Brain and Immune Function produced by Mindfulness Meditation, Psychosomatic Medicine 65:564–570, 2003.
13. Lustman PJ and Clouse RE, Depression in diabetic patients: the relationship between mood and glycemic control, Journal of diabetes and its complications. 19:113–122, 2005.
14. Hankey A, Studies of Advanced Stages of Meditation in the Tibetan Buddhist and Vedic Tradition. I: A Comparison of General Changes, *eCAM* 2006; 3(4)513–521.
15. Ott MJ et al, Mindfulness Meditation for Oncology Patients: A Discussion and Critical Review, integrative cancer therapies 5(2):98–108, 2006.
16. Cheng RWF et.al, Human prefrontal cortical response to the meditative state: A spectroscopy study: J. of Int. Neuroscience,120, 483-488, 2010

LaVergne, TN USA
15 March 2011
220073LV00007B/7/P

LV000207218 56